SCIENCE, PSEUDO-SCIENCE
AND SOCIETY

Essays by

Paul Thagard
Adolf Grünbaum
Antony Flew
Robert G. Weyant
Marsha P. Hanen
Richard S. Westfall

Trevor H. Levere
A. B. McKillop
James R. Jacob
Roger Cooter
Margaret J. Osler
Marx W. Wartofsky

Edited by

Marsha P. Hanen, Margaret J. Osler,
and Robert G. Weyant

Published for
The Calgary Institute for the Humanities
by
Wilfrid Laurier University Press

Canadian Cataloguing in Publication Data

Main entry under title:

Science, pseudo-science and society

Papers presented at a conference sponsored by the Calgary Institute for the Humanities, and held at the University of Calgary, May 10-12, 1979.

ISBN 0-88920-100-5 pa.

1. Science − Philosophy − Congresses. 2. Science − History − Congresses. I. Hanen, Marsha P., 1936- II. Osler, Margaret J., 1942-   III. Weyant, Robert G., 1933-   IV. Calgary Institute for the Humanities.

Q174.S44   501   C80-094537-9

Copyright © 1980
Wilfrid Laurier University Press
Waterloo, Ontario, Canada
N2L 3C5
80 81 82 83 4 3 2 1

TABLE OF CONTENTS

iv

FOREWORD

This volume collects the papers presented at a conference on "Science, Pseudo-science and Society," sponsored by the Calgary Institute for the Humanities and held at the University of Calgary, May 10-12, 1979. More than many such collections, this one preserves some trace of the intellectual excitement which surrounded this gathering of scholars. All papers delivered at the conference have found their way into this volume, except that by James Jacob and Margaret Jacob, "The Anglican Origins of Modern Science: The Metaphysical Foundations of the Whig Constitution," for which the Institute, reluctantly, had to yield to the prior claim of ISIS, which will publish the paper in Summer, 1980.

The conference was the fourth such event sponsored by the Calgary Institute for the Humanities as part of its function to augment and enrich advanced study and research in the humanities.

We wish to express our gratitude to those who arranged the meetings: Terence Penelhum, the Institute's first Director; Marsha Hanen, Margaret Osler, and Robert Weyant, the co-chairmen of the conference and editors of this volume; and to Mrs. Lois Kokoski of the University's Conference Office who shouldered more organizational details than can be mentioned here. Penny Williams taped the entire proceedings for use on the CBC "Ideas" program, and her efforts to interview groups of participants on key issues added considerably to dialogue at the symposium.

For grants in aid of the conference we wish to express thanks to the Social Sciences and Humanities Research Council of Canada, the University of Calgary Research Policy and Grants Committee, as well as to the Faculty of Social Sciences, the University College, and the Department of Psychology, all of the University of Calgary.

We are especially grateful to the volume editors for assembling and preparing the text, and to Gerry Dyer for the difficult task of preparing the final manuscript of this book.

Egmont Lee,
Director

ABOUT THE AUTHORS

Roger Cooter is presently a Killam Post-Doctoral Fellow at Dalhousie
University, and was Post-Doctoral Fellow at the Calgary Institute
for the Humanities in 1979-80. A social historian, he received
his B.A. at Simon Fraser University, his M.A. from the University
of Durham, and his Ph.D. from the University of Cambridge. He is
the author of *The Cultural Meaning of Popular Science: Phrenology
and the Organization of Consent in Nineteenth Century Britain*, to
be published by Cambridge University Press. He has also written
articles on British social history and the social context of science.

Antony Flew, currently Professor of Philosophy at the University of
Reading, received his B.A. and M.A. from Oxford and his D.Litt.
from the University of Keele. He has published widely in philosophy, with a special interest in the philosophical problems of
religion and psychical research as well as the philosophy of the
social sciences. His books include *A New Approach to Psychical
Research*, *God and Philosophy*, and *Sociology, Equality, and Education*, as well as the recent *Rational Animal*.

Adolf Grünbaum, Andrew Mellon Professor of Philosophy at the University
of Pittsburgh, Chairman of the Centre for Philosophy of Science,
and Research Professor of Psychiatry at that University received
his B.A. from Wesleyan University and his M.S. and Ph.D. from
Yale. He is the author of *Philosophical Problems of Space and
Time*, *Modern Science and Zeno's Paradoxes*, and *Geometry and
Chronometry in Philosophical Perspective* as well as a large number of influential articles in philosophy of science. He has been
President of the Philosophy of Science Association; and he has
also been appointed an Einstein Centennial Lecturer under the
auspices of the Institute for Advanced Study in Princeton. A new
book, *Is Psychoanalysis a Pseudo-Science?* will be published soon
by the University of California Press.

Marsha P. Hanen, Associate Professor of Philosophy and Associate Pro-
vost of University College at the University of Calgary, special-
izes in the philosophy of the natural and social sciences and the
philosophy of law. She received her B.A. and M.A. from Brown
University and her Ph.D. from Brandeis University, and recently
spent a year as a Fellow in Law and Philosophy at Harvard University. She has published articles on aspects of the philosophy of
science and law and has spoken widely in England and North America.

James R. Jacob, Associate Professor of History at John Jay College of
the City University of New York, received his B.A. from the University of Tulsa, his M.A. from Rice University, and his Ph.D.
from Cornell. Specializing in the social dimensions of English
science in the seventeenth century, he is the author of *Robert
Boyle and the English Revolution*, as well as articles on science,
society, and religion in seventeenth-century England.

Trevor H. Levere, Associate Professor at the Institute for the History
and Philosophy of Science and Technology at the University of
Toronto, received his B.A., M.A. and D.Phil. from New College,
Oxford. His numerous publications include *Affinity and Matter*
and the forthcoming *S. T. Coleridge and Early Nineteenth-century
Science*. He has published many articles on the history of chemistry in the nineteenth century.

A. B. McKillop, Assistant Professor of History at the University of
Manitoba, received his B.A. and M.A. from the University of Mani-
toba and his Ph.D. from Queen's University at Kingston, Ontario.
He has published articles on Canadian intellectual history, as
well as *William Dawson Le Sueur* and the forthcoming *A Disciplined
Intelligence: The Development of Critical Enquiry in Anglo-
Canadian Thought*.

Margaret J. Osler, Associate Professor of History at the University of
Calgary, received her B.A. from Swarthmore College and her M.A.
and Ph.D. from Indiana University. She has published a number of
articles on the mechanical philosophy and science in the seventeenth century.

Paul Thagard, Assistant Professor of Philosophy at the University of
Michigan-Dearborn, did his undergraduate work at the University
of Saskatchewan and Cambridge University, received his M.A. from
Cambridge and his M.A. and Ph.D. from the University of Toronto.
He has published a number of articles on various aspects of the
philosophy of science.

Marx W. Wartofsky, Professor of Philosophy at Boston University and
Director of the Boston Colloquium for the Philosophy of Science,
received his A.B., M.A., and Ph.D. from Columbia University. He
is author of *Conceptual Foundations of Scientific Thought*,
Feuerbach, and the forthcoming *Models: Representation and the
Scientific Understanding*. He is co-editor of the *Boston Studies
in the Philosophy of Science*.

Richard S. Westfall is Distinguished Professor of History and Philosophy of Science at Indiana University. He received his B.A., M.A.,
and Ph.D. from Yale. He is a specialist on science in the seventeenth century, particularly the work of Isaac Newton. His numerous publications include *Science and Religion in Seventeenth
Century England* and *Force in Newton's Physics*, as well as a
forthcoming biography of Newton.

Robert G. Weyant, Professor of Psychology at the University of Calgary
and former Dean of the Faculty of Arts and Science, received his
B.A. from Lafayette College, his M.A. from Kent State University,
and his Ph.D. from the University of Iowa. His area of research
is the history of psychology, especially in the eighteenth century. He is editor of a facsimile edition of Condillac's *Essay
on the Origin of Human Knowledge* and has published a number of
articles on the history of psychology.

PREFACE

A primary inspiration for the symposium on "Science, Pseudo-science, and Society," which took place at the University of Calgary on May 10, 11, and 12, 1979, was a growing awareness of the crucial role the study of pseudo-science plays in the areas of contemporary scholarship which are concerned with the nature of science and its relationship to broader social issues.

In the recent past when varieties of logical positivism dominated the philosophy of science, a central goal was to formulate a criterion of demarcation that would decisively separate science from every other area of intellectual activity. During the past two decades, developments in social history, history of science, and philosophy of science have undermined the hopes of earlier generations of positivists. Indeed, some historically informed philosophers of science have come to question the very possibility of developing formal criteria for an adequate definition of science. Whereas the earlier generation of philosophers of science typically wanted to draw a sharp line between the context of discovery and the context of justification and seriously debated whether history and philosophy of science had any bearing on each other, recent developments indicate the absence of any sharp separation between these areas. Indeed many contemporary scholars would argue that the areas of overlap and mutual penetration are so great that such distinctions of content and field are no longer possible.

Recent scholarship, partly as a result of the counterculture's fascination with astrology and the popularization of various occult subjects, has turned to an examination of pseudo-science and its role in intellectual and social life. Pseudo-science purports to be scientific and yet is denied that status by practitioners of science. Thus, understanding pseudo-science becomes central to understanding what the scientific community regards as properly scientific. At the far end of the spectrum, historical relativists claim that science is simply what the community of scientists at any given time decides to call science. Others temper the demands of historical context with the underlying assumption that science does possess distinct, recognizable, and specifiable features. All agree that science is one of the most

important developments of the last four hundred years of European and
North American civilization and that it is a phenomenon worthy of serious study.

Although most people would readily ascribe the name "science" to
astronomy, chemistry, and studies of the localization of brain function
and the name "pseudo-science" to astrology, alchemy, and phrenology,
the task of formulating a clear, explicit demarcation between them is
not straightforward. For one thing, the designation of a field as one
or the other is not a once-and-for-all, atemporally valid decision.
Rather, the designations vary with historical context. Some scholars
would even argue that the designation involves the political, economic,
and social uses of the fields in question. In the present volume, Paul
Thagard, Antony Flew, and Adolf Grünbaum hold out for the possibility
of a formal demarcation between science and pseudo-science, while James
Jacob, Robert G. Weyant, and Roger Cooter argue that the demarcation is
determined by social, political, and historical factors. Although this
latter thesis is controversial, no one doubts that the pseudo-sciences
so-called have exerted a formative influence on the areas that later
came to be recognized unequivocally as sciences. Richard S. Westfall
argues that Newton's concept of force may well have been inspired by
his alchemical studies, just as, according to Trevor Levere, Coleridge's
concern with the pseudo-sciences of his day pertained to his conception
of science.

Pseudo-science, recognized if not defined, is pivotal to understanding some of the historical relationships between science and society. Pseudo-science is often the vehicle by which science and the
scientific mentality are popularized. If, as some would argue, science
has attained a priestly role in contemporary society, it has done so
more often through its imitators than its practitioners. Not surprisingly, astrology, armchair anthropology, and science fiction are more
familiar to the general reading public than the arcane principles of
quantum mechanics. Not only has pseudo-science often served as the
popularizer of scientific thinking, it has thereby been used - consciously or unconsciously - to influence popular thought. Or so the
more radical historians would have us believe.

This volume is organized around three major questions concerning
the relationships among science, pseudo-science, and society. The
papers in the first section address the question of whether it is
possible to draw a sharp demarcation between science and pseudo-science

and what the criteria of that demarcation might be. The papers in the
second section, recognizing the historical importance of various of
the pseudo-sciences, consider their impact - positive or negative - on
the development of the sciences themselves. The papers in the third
section deal with the question of the relationship between the sciences
and pseudo-sciences, on the one hand, and social factors on the other.
Marx Wartofsky, who performed with grace the difficult task of providing ongoing commentary for the entire symposium, has written an introduction and final commentary for the book.

INTRODUCTORY REMARKS

It is my lot, if not my duty, in presenting these opening remarks
at our conference, to take the title of our meeting seriously. That,
it turns out, is not such a bad thing. I understand that the writer
William Saroyan would invent various titles, and when he hit upon one
he liked, he would write a short story for it. I like our title:
Science, Pseudo-Science and Society - not least, because it manages to
alliterate the "s" sound, both at word beginnings, and internally, no
fewer than seven times in three words. It is a sumptuously sibilant
conference title, and ranks with the best of Nelson Goodman's efforts.
I also like the title because it is trinitarian, or at least triune.
Short of being outrightly theological, it hints at dialectic thirds,
at Peirce's theory of meaning, at internal relations; surely a rich
setting for dialogue and debate.

More seriously, however - (not *much* more seriously) - these titu-
lar musings lead me to another methodological point: does the title
determine the content of the conference, or serve as a heuristic framework for the papers to be presented? Or do the contents, in their
variety, inductively generate or give evidential support for the title?
Here, the issue is joined. Are conference titles like theories in
science? Or like research programs? Will we, in fact, talk about
science and pseudo-science, and about their social contexts? Can we,
may we, in a weakly deductive way to be sure, make the inference that
the papers and the discussion will instantiate the general conceptions,
the thematic suggestions which the title represents? And if so, what
are those conceptions and suggestions? As a colleague of mine likes to
say: "What's the problem?"

Now, finally, I am forced to be fully serious. This is to warn
you that what I will say now is no joke. It is, indeed, an anticipa-
tory comment, and a prospective critique of what will come to be said
here. The problem, as I see it, is to distinguish pseudo-science,
proto-science and science proper. But this is no mere definitional
problem, nor even a demarcation problem *tout court*. Demarcation problems were easy: There was science, and there was metaphysics. The job
was to get past metaphysics to science, having already gotten past
theology and other forms of mythology. Auguste Comte and Ludwig

Feuerbach had, independently of each other, and at about the same
time, come up with the classic division of the three stages of human
thought: theological, metaphysical, scientific; and whatever else
they meant by "scientific", science was "*positive* science", it was empirical, and it was true. The rest was speculation or myth, either
degenerate, or at best adumbrative, not yet face to face with reality,
but presenting it, representing it, or hiding it under the misty veil
of feeling, poetic fancy, abstract speculation.

Hume had earlier demarcated in no uncertain terms, in an inquisi-
torial passage consigning to the flames what fell outside of "matters
of fact and ratiocination." *We* inherited all this by way of J. S. Mill,
C. S. Peirce, E. Mach. If Kant had only had the good sense not to mess
around with the synthetic *a priori*, and matters of teleology and aesthetic judgement; if he had stayed away from regulative ideas, he too
would have come with clean hands to the demarcation game. With our
Viennese forebears and their German colleagues, the demarcation lines
were drawn hard and fast. And then erased and redrawn. And then
fuzzed out and erased again. And redrawn again. Reduction to the empirical basis, verificationism, operationism, falsificationism, naive
and sophisticated, each had its moment of clarity, of triumph, only to
fall to the successive waves of criticism, internal and external, which
washed away the demarcation lines like so many inscriptions in the
sand.

What had started as an historical critique and a developmental
thesis about the genesis and unfolding of ever higher stages of human
knowledge ended as an analytical-logical reconstruction of "science"
as it was ostensibly done by proper scientists, and as it was articu-
lated in the set of statements which constituted the body of knowledge
which was science proper. Which scientists were proper and which
statements were proper it was the task of the demarcation criterion
(or criteria) to identify. The pendulum-swing from genetic-historical
to analytical and back to genetic-historical has taken place in the
philosophy of science in the last century. But the pendulum sweeps out
no ideal Galilean arc: it wobbles, it has oscillations imposed upon
oscillations to produce a very complex wave-function indeed. Right
now, the historical is in the ascendant, to be sure. But a half-
century of logical reconstruction and analysis is not to be so easily
cancelled out. The new historicism, the recent concern with scientific change and the growth of knowledge, variously called

post-positivism, or "the new philosophy of science", has opened a
Pandora's box of conceptual demons, threatening the neatly ordered
structures of earlier logical-analytic law and order. What is put in
question is precisely the demarcation game itself. All the earlier
demonology, which had developed appropriate instruments and rituals of
exorcism, no longer suffices. The devil of metaphysics used to yield
to the sign of the cross - i.e. to the exorcism of the crucial experiment, to the criterion of empirical testability, or falsification. No
longer.

The imp of poetic imagination, the demon of metaphor, the fallen
angel, Lucifer, with his theological-mystical penchant for interpreting
science as an emanation from the Godhead, or from number - all of these
vanquished, exiled, exorcised contaminators of scientific rationality
are back; in mutant forms to be sure, but back nevertheless. And how?
In two ways: first, resurrected by the historians and sociologists of
science, who discover them lurking in the very innards of scientific
thought, in its actual historical practice. Thus, it is not only
Feyerabend, the wild man of Berkeley and Berlin, with his voodoo
chants, who insists that it is the crazies who make science go. As
sober and inductivist a historian as Stillman Drake calls our attention
to the role of music in Galileo's thought. We used to be able to write
off the neo-Platonism or the number-mysticism or the dream-work which
figured in the work of Kepler, Copernicus, Galileo, Newton, Descartes,
Leibniz - we could attribute it to the vestigial influences of earlier,
pre-scientific thought. Now we are told that these very ideas were
creative, constitutive components of the scientific revolution itself.
The devils have worked their way back to a new legitimacy. They bore
from within, and get assigned establishment jobs.

The second way in which the devils return is perhaps even more
serious: scientists are people. Their beliefs, their wishes, their
interests, their social and class position in society, their hang-ups
and their hang-ons all play a role in shaping their thought, in focus-
ing their attention, in drawing their loyalties and their resentments,
their hopes and their fears. Scientists, like the rest of us, have
mothers and fathers, and we all know what *that* does to the psyche!
(If we don't, Grünbaum will remind us what the Freudian fashions are
in this regard.) So the social, the psychological, the historical,
the economic come to be seen as contexts, if not causal factors, of
scientific thought and change. What Lakatos used to like to call

"externalist rubbish" is at least taken to be seriously problematic in
the contemporary historical reconstructions of science. Arguments
about the genetic fallacy, the fact-value distinction, and the difference between causes and reasons used to be useful weapons against such
externalist devils. No longer. The mutant forms are resistant to the
older antibiotics, to mix a metaphor. Or two. Or three.

Thus, if metaphysics and metaphor, on the one hand, and social,
political and psychological contexts, on the other, now enter into the
historical reconstruction of science, if the philosophy of science is
bedeviled by these intrusions, then the old demarcation game is over.
The simple divisions of an earlier time will no longer do. Yet, does
anything go?

This conference proposes to deal with science and pseudo-science;
therefore, not yet with science and metaphysics, or science and religion, nor yet with science and society. Pseudo-science is not merely
what is *not* science proper. It is not to be defined by exhaustive
disjunction, or by some method of residues. Rather, it is what either
appears *as* science, or represents itself *as* science. Or else, it is
what science proper - the scientific establishment, or the scientific
inquisition - marks off as heretical: not simply alternative, or
alien, or other but a clear violation of the true faith. What is dialectical about such a formulation is that pseudo-science cannot be
determined as such without at the same time determining what is properly scientific. These are interdependent concepts. But without as yet
determining what is scientific and what pseudo-scientific, in a substantive way (which is the problematique of the conference papers and
the discussion itself), one may still make some formal and conceptual
distinctions. For example, if we begin with the presupposition that
there is such a thing as pseudo-science, then this bears with it an
implicit epistemological assumption - namely, that pseudo-science is in
some sense *false*, or that it pretends to achieve what it cannot
achieve; and, conversely, that science gives us truth, or delivers what
is expected of it. Thus (whether one begins from a realist or a prag-
matist theory of truth), this distinction between the real thing and a
fake is clearly supposed. Enlightened rationality demands that we root
out falsehood, fakery and pretense. Even so weak and piddly a method
as ordinary language analysis can yield the conclusion that the false,
the fake and the pretentious ought not, as we understand and use these
terms, to command our intellectual respect, nor be taken to be

constitutive factors in forming our beliefs. Therefore, whatever we
define as pseudo-science will be *bad stuff*. If we start with the premise that *pseudo* is *bad*, then of course the morality of the intellect
demands that we demarcate once again, hunt down the fakers, and eliminate them root and branch so that we can get on with the search for
truth unobstructed.

Now let us look at our conference program to see how pseudo-
science (or pseudo-scientific reasoning) is characterized, what sort of
role it plays *vis-à-vis* science proper, how it is to be understood historically, and in the present. As we will see, when the conference un-
folds, there is a wide spectrum of alternative approaches, which make
any simple demarcation of science from pseudo-science untenable.
Thagard's paper is neat: it takes the paradigmatic *pseudo*-science,
astrology, and the paradigmatic *science*-science, physics, and examines
in what way they are essentially different. But does (or did) astrology (as pseudo-science) play a role in the development of science-
proper? Grünbaum attacks psychological explanations, in particular
Freudian ones, as reasons for accepting or rejecting scientific theories. He characterizes and criticizes such reasoning as pseudo-scientific. Westfall talks of Newton's alchemical experiments and the large
but unacknowledged role they play in his scientific thought. Is Newton's science thereby tainted, or not? James Jacob reconstructs
Stubbe's critique of the Royal Society as a radical critique of establishment religion. The title of the joint paper by James and Margaret
Jacob is a thesis: "Anglican Origins of Modern Science."[1] So politics
and theology are certainly seen as part and parcel of the scientific
game. Antony Flew takes a sustained and sober look at parapsychology.
Pseudo-science or proto-science? Trevor Levere brings Coleridge's
critique to bear upon the human sciences and deals with phrenology and
mesmerism. McKillop introduces poetic fancy in the *reception* of scientific theory. Roger Cooter proposes that "pseudo science" has served
historically, and continues to serve as a label to preserve the *status
quo ante* of establishment science. The "shock-value" of his thesis is
that even radicals in science "deploy" this label to ascribe ideological motives only to the "pseudos", thereby covering up in positivist
fashion once again the fact that *all* science is ideological. Margaret
Osler addresses the use of appeals to "scientific facts" by opposed
ideologies, which, together, draw contradictory conclusions from such

facts. She attacks the mode of reasoning which uses science to justify ethical or social problems.

Strange goings on. For on the one hand, we have arguments for
the *active* role, even the constructive role of theology, politics,
literature or literary theory, alchemy, within science proper; on the
other, we have astrology, Freudian theory, phrenology, mesmerism, parapsychology as (prospectively) diversions from, subversions of, or pre-
tenses at science proper, or at best as problematic claimants to the
status of sciences. Are these to be understood as pseudo-sciences or
proto-sciences? Are we going to say that pseudo-science, as fakery,
falsehood and pretense, plays a role in the growth of science *proper*?
Then it isn't such *bad stuff*. Or are we going to say instead that
pseudo-science is destructive to scientific thought, and to the reception and public understanding of science? Further: Is pseudo-science
a disease which affects only the non-scientific public, and to which
scientists are immune? Or are proper scientists (like Newton) capable
of being infected as well? For infectious diseases there are three
sorts of remedies: Curative therapy, preventive therapy and quarantine. The first is for those already infected; the second and third,
for the sake of those not yet exposed. How shall we then deal with
pseudo-science, if it is indeed a Bad Thing? Shall we inhibit its
spread by quarantine, i.e. by censorship, by keeping pseudo-science
from being taught and publicized? By executing the leaders? By ino-
culating the as-yet uninfected populace with antibodies (anti-thoughts)?
But what if what is taken to be *pseudo*-science is in fact *proto*-science?
Or better yet, what if so-called pseudo-science is the inevitable form
which a science proper takes in its earliest conjectural and innovative
stage? To cut if off at *that* point is to kill the bud and forever
forestall the flower.

Popper argued both for a strong demarcation, in terms of contexts
of justification (or contexts of falsification - it all comes to the
same thing), and for *no* demarcation in contexts of discovery. For
Popper, the sources of scientific ideas fell beyond the demarcation
injunction. Conjecture had no "logic." But the demarcation of what
was scientific from what was not scientific could, he argued, be
established by logical means: Any universal (law-like) statement which
could not potentially be falsified by some conceivable, empirically
testable singular statement, was, by virtue of this unfalsifiability,
not a scientific statement. The logic of denying the consequent, in

Modus Tollens, was the requisite form which yielded the criterion of
science proper an empirical interpretation. That demarcation, I think,
doesn't work anymore. It has been shredded by the post-and-anti
Popperians. The task of this conference, then, is not simply to be
able to determine for us what we will all regard (or already regard) as
pseudo-science as against what we regard as science. Rather, the task
is to make explicit the terms in which, if there are demarcations to be
made, they can now be made, *in view of the failure of past criteria of
demarcation*.

Beyond the relation between science and pseudo-science, there is
one other context which the title suggests, and which it is reasonable
to expect that our conference discussion will pursue, namely how does
society, or the social context of science play a role in the relation
between science and pseudo-science? Here we face a choice between
Scylla and Charybdis. On the one hand, one may say that science and
pseudo-science are divided by some essential difference between them,
either in method, or in logical form, or on epistemological grounds.
That's Scylla. Or, charybdismatically, we could say that it is really
a matter of social determination or social choice as to what constitutes science proper and what constitutes pseudo-science, that this is
ultimately an historically and culturally relative, or socially relative question. There are antecedents for such relativism in other do-
mains of human thought and practice. Why should science, as a human
activity and as a social institution, be exempt from such an understanding? So for example, disease has been defined as whatever it is
that doctors treat. The entity varies with changes in medical theory
and practice. The ontology of disease is therefore shifty, variable,
institutional. Analogously, in André Malraux's phrase, art is what is
in the museum. Or again, in Voltaire's version, religion is an invention of the priests. And therefore, the argument goes, science is
what's taken to be science at any given time and place by the social
institution of science, whereas, pseudo-science is whatever the establishment or the culture takes to be pseudo-science. There are no in-
trinsic logical, epistemological or other essential criteria which
themselves escape this relativism. Thus, between the Scylla of Platon-
istic essentialism and the Charybdis of historical and social relativism, the question of demarcation then becomes a much more complex
problematic than it has been in the past, once the relativist demons
have been let out of the box. (It is unnecessary, with this audience,

to give the names of the various Pandoras who are responsible for our
present troubles. Besides which, some of my best friends are relativists; and, historically speaking, consorting with demons, or with
demon-releasers is a dangerous practice back home in Massachusetts, and
I don't want to get into trouble.)

The one last thing I want to say is that in distinguishing science
from pseudo-science, or both from what is proto-science, we may choose
to adopt a method of intuition and simply say, "Well, we know in advance, or we know somehow, or we're capable of making subtle and well-
grounded judgements about these things, without at the same time knowing *how* we do this. We are able to do chicken-sexing, and we are able
to tell good from bad violin-playing, without yet having any explicit
theory as to how we are able to make these determinations. That doesn't
make the determinations less valid, or less persuasive." Yet, it does
seem to put the burden on us for explaining how we make such determinations, and what it is that gives our ineffable intuitions in these matters such force. St. Augustine's famous statement about the nature of
time is relevant here. He says, "If you don't ask me what time is, I
know; but if you ask me, then I don't know." I think what we are up
against here is being asked what are the criteria by which we make the
intuitively valid distinctions between science and pseudo-science, and
do these criteria preserve the kinds of distinctions we want to maintain? The critical function of asking the question as to what the criteria are is presumably to overcome whatever dogmatism there is in our
present intuitions. The critical function of the demarcation game is
not to establish once and for all, to all of our satisfactions, some
algorithm by means of which we can then make the decision, or arrive at
some decision procedure by which we can neatly separate science from
pseudo-science. The critical function of the demarcation discussion is
to make us subtly and sensitively aware of the difficulties of simply
enforcing our present intuitions, and to hold us responsible for arti-
culating what the grounds are for our judgements. This leaves us, at
the very least, demarcational fallibilists, and I hope that's where we
end up, because that's where I am, and I believe I'm right.

FOOTNOTES

1 This paper, although presented at the conference, was not available
   for publication in this volume. It is being published in *ISIS*,
   June 1980.

I. THE PROBLEM OF DEMARCATION

The papers in this section are all concerned in some way with demarcation. All of the authors represented here believe that we can differentiate between science and pseudo-science, though not necessarily by means of a clear-cut criterion that will give us unequivocal answers in every case. Indeed, it is fair to say that none of these authors is putting forward a criterion of demarcation in the old style, though several of them offer us guidelines of various kinds for drawing the distinction. We also find in this section varying degrees of emphasis on a tie to history: no-one is any longer doing ahistorical philosophy of science, but it is clear that a number of these authors hold that there are normative considerations over and above the purely descriptive ones in the question of what is and what is not science, and that it is incumbent on philosophers to bring these forward.

Paul Thagard addresses directly one aspect of the conference theme by suggesting and exploring a criterion for distinguishing between science and pseudo-science: the former uses correlation-thinking whereas the latter depends upon resemblance-thinking. The crucial feature of resemblance-thinking is not that it appeals to similarities, but that it infers causal relations from resemblances. This is what is unscientific - not the use of resemblances *tout court*, for, Thagard insists, there is a perfectly legitimate use of similarities in the form of analogies and metaphors in scientific thinking as in any other. But the use of resemblance-thinking does not serve to draw a distinction between science and pseudo-science: resemblance-thinking (in Thagard's sense), it turns out, is sufficient but not necessary for a field to be pseudo-scientific.

We must, however, add the caveat that not all resemblance-thinking in the broader sense is pseudo-scientific, for resemblances are used in ways that do not purport to be scientific, and in areas where causal questions are not at issue, as sometimes in literary interpretation. Pseudo-science is characterized partly by the use of resemblance-thinking, but also by unprogressiveness, and by an absence of critical evaluation in the face of problems.

Adolf Grünbaum asks what is and should be the role, if any, of psychological explanations in the acceptance or rejection of scientific theories. He deplores "dismissive" psychological explanations of the sort that look not to a theory's credentials in the form of evidence for or against, but to the psychological motivations of its author. Using psychoanalytic theory as an example Grünbaum insists on a sharp differentiation between *evidential* warrant for a theory and acceptance or rejection of it on emotional grounds; and, he says, to argue that any rejection of a given theory must be based on neurotic resistance is obviously question-begging. But emotional attraction to or revulsion against a theory may have heuristic benefits in inspiring the unearthing of objectively favorable or contrary evidence.

As to the propriety of inquiring into psychological motivations for accepting or rejection of some theory, Grünbaum tells us that we must first consider the weight of the evidence; for both true and false beliefs are caused, and the difference between them lies not in their source but in their evidential warrant. And we need to be especially careful when the psychologistic dismissal occurs at, as it were, the second level - where, that is, it is claimed that the theory's proponent is only pretending to consider relevant evidence, but that his espousal of it can only really be explained on the basis of some idiosyncratically intense personal preoccupation. In the case of such a claim, it is especially important that we look to the evidence offered for the theory to see whether it, rather than some extra-evidential material, might indeed be the basis of the theory's acceptance. So, according to Grünbaum, although motivational considerations often explain why someone accepts (or rejects) a given theory, it would be a mistake to think that such factors can help us to adjudicate the scientific merits of the theory in question.

Antony Flew takes up anew[1] the question of parapsychology, pointing out first how it differs from other cases such as astrological prediction, UFO's, or the alleged mysteries of the Bermuda Triangle or the Chariots of the Gods. But also, parapsychology differs significantly from the high status sciences. Drawing a distinction between two kinds of psychic phenomena - psi gamma (clairvoyance and telepathy) and psi-kappa (psychokinesis), Professor Flew argues that the first, at least, cannot reasonably be characterized as a new form of either perception or knowledge. There are a number of important similarities between the two kinds of psychic phenomena including the fact that in both

cases there seems to be an absence of repeatability. Now, no doubt
repeatability is always important in scientific investigation but,
Flew argues, it is particularly crucial in connection with parapsychology because this area has seen so much fraud and self-deception. Without repeatability, then, parapsychology is open to the Humean challenge
against miracles - a challenge it is not in a very good position to
meet.

A further problem with psi-phenomena is that "there is no even
half-way plausible theory with which to account for the material it is
supposed to have to explain," even after a hundred years of working
with these materials. And, in addition, it appears that some of the
phenomena in question "are so defined as to be necessarily impervious
to causal explanation," in a way that ordinary phenomena are not, even
if we do not at the moment have that explanation, and even if our current theories seem unable to provide one. Thus, although it would
have been wrong to dismiss out of hand the claims of parapsychology to
scientific status, Flew would presumably say that there has now been
enough investigation of these matters without favorable results for us
fairly to apply the label "pseudo-science."

Robert Weyant's paper begins with an attempt to clarify some of
the concepts central to the science/pseudo-science distinction. The
major distinction, he tells us, is not that between science and pseudo-
science, but rather that between science and non-science; and it is to
this distinction that traditional criteria of demarcation have been
addressed. In the former realm we find untested but scientific material, as well as "correct" and "incorrect" science. In the realm of
non-science, there is that which lays claim to being science, and that
which does not, such as literature, art appreciation, or theology.
Claims to scientific status are made both by potential or protosciences
and by pseudo-sciences; and Professor Weyant focuses in this paper on
the judgment between these two.

Turning next to metaphorical thought, Weyant accepts the views of
Black and Hesse, according to which metaphors are not mere ornaments
to thought, but are rather aids to understanding, not in the least
devoid of cognitive content. The case of Mesmer provides an interesting example, for here we have the use of metaphors in aid of an incorrect theory, and indeed one often taken to be fraudulent. And yet it
is not immediately clear what differentiates Mesmer's use of metaphorical thinking from that of Galileo, Bacon, Gilbert, Newton, or Boyle.

If it is that controlled tests show Mesmer's theory of animal magnetism
to be false, then it would appear that we are classifying the theory as
"incorrect" science, and thus not as pseudo-science. But, Weyant
claims, Mesmer's theory is really not testable at all (because it provides an explanation for any conceivable outcome of any conceivable
empirical test) and indeed it is even internally inconsistent, so it
can only appropriately be viewed as pseudo-scientific.

But we should not suppose that the judgment that something is a
pseudo-science is absolute, or purely logical, or dependent exclusively
on empirical evidence. Instead, we must be aware that there are political, social, economic, theological, and professional factors that
enter into such judgments, and our decisions about appropriate classifications may well vary from time to time and place to place, depending
upon these factors.

Marsha Hanen's paper addresses the demarcation problem from yet
another perspective - this time the question of what kind of line, if
any, can be drawn between science on the one hand and certain other
social pursuits, especially law, on the other.

Starting from Ronald Dworkin's coherence theory of judicial justification, Professor Hanen points to analogies between that theory and
coherence theories of justification in philosophy, in linguistics, and
in science. She argues that the two models for coherence - the natural
and the constructive - that Dworkin puts forward are more similar than
he supposes, and that the appropriate model for both science and law is
neither of these but rather some composite of the two. In addition,
she maintains that the position that advocates two separate models
rests on a too strongly realist picture of science, and she offers an
alternative conception using examples from the history of science. In
particular, she argues that Dworkin's contention that the constructive
model but not the natural allows that we simply disregard recalcitrant
intuitions or observations that do not fit the best available theory is
too strong: that insofar as such ignoring is appropriate, it is so as
much for science as for law and morals; but that it is rare in all
spheres, for we nearly always feel bound to keep trying to reconcile
theory and data.

In the latter part of the paper these considerations are applied
to the problem of the standard of care required of police officers in
apprehending a suspected criminal, and in particular the extent to
which police officers are justified in using force in effecting an

arrest. The two cases considered raise questions about the application
of precedent, and especially about the notion of mistakes and how these
can be corrected. In both cases there were injuries which occurred
during a police pursuit, and there were enough similarities between the
two situations that one might expect plaintiffs to recover or fail to
recover in both cases. On the other hand, there were features according to which one might think that the court could reasonably apply a
legal doctrine that held the first officer liable and exonerated the
second. But the actual result was, oddly, the reverse, and Hanen considers whether any plausible principled justification of this result
can be made so as to yield a coherent theory in at least this small
corner of law. She suggests that, although a coherence theory applies
to justification in both law and science, the two do differ importantly
in content, legal justification requiring an appeal to moral considerations not usual in justifying scientific theories.

FOOTNOTES

1 Professor Flew has written extensively on this topic over the past
   twenty-five years, beginning with his *A New Approach to Psychical
   Research*, London: C. A. Watts, 1953. See the footnotes to his
   paper for further references.

RESEMBLANCE, CORRELATION, AND PSEUDOSCIENCE

I want to argue that there is a fundamental distinction between
two kinds of reasoning, which I call resemblance thinking and correlation thinking. Resemblance thinking infers that two things or events
are causally related from the fact that they are *similar* to each other;
in contrast, correlation thinking infers that two things or events are
causally related from the fact that they are *correlated* with each other.
After presenting psychological, anthropological, and historical evidence for the importance of the distinction between resemblance and
correlation thinking, I consider whether that distinction provides the
basis for the distinction between pseudoscience and science. It turns
out that the distinctions do not completely mesh: I argue that the use
of resemblance thinking is sufficient but not necessary for branding as
pseudoscientific such disciplines as astrology.

In his *System of Logic*, Mill identified as one of the most "deeply-
rooted" of all fallacies the belief "that the conditions of a phenomenon must, or at least probably will, resemble the phenomenon itself."[1]
Suppose you are asked to judge whether people with red hair are generally hot-tempered. The correlational approach would be to take a sample
and count the numbers falling under the various categories of having or
not having the two properties in question. In fact, most people would
merely summon to mind a few examples of red-headed acquaintances and
thereupon tender a judgment. Or, using resemblance thinking, they
might notice a similarity between the fiery appearance of red hair and
the metaphorically fiery behaviour of hot-tempered people, and use this
to judge that red-heads are in general hot-tempered.

As was pointed out by Mill, and by Lawrence Jerome in his book,
Astrology Disproved, astrology is rife with resemblance thinking. Jerome describes how astrology is based on a "principle of correspondences"
or "law of analogies."[2] For example, the reddish cast of the planet
Mars leads to its association with blood, war, and aggression, while
the pretty "star" Venus is associated with beauty and motherhood.
Saturn, which is duller and slower than Jupiter, is associated with
gloom, and (one hopes even more tenuously) with scholarship. Similarly, analogy attaches characteristics to the signs of the Zodiac: Libra,
represented by the scales, signifies the just and harmonious, while

Scorpio resembles its namesake in being secretive and aggressive. The
associations concerning the planets and signs of the Zodiac are taken
by astrologers as evidence of some causal influence of the heavens on
the personalities and fates of individuals whose births occur at the
appropriate times. Until very recently, no attempt was made to determine whether there is any actual correlation between the characteristics of the signs and planets and the personalities of the people under
their alleged influence.[3]

I want to argue that resemblance thinking is much more than just a
quirk of astrologers. Rather, it is a method which is very natural for
human beings in general. The first piece of evidence for this is found
in the important psychological investigations of Amos Tversky and Daniel
Kahneman. According to Tversky and Kahneman, people making judgments
about the relation of classes or events typically make use of a "repre-
sentativeness heuristic," basing their judgments on degree of resemblance.[4] Suppose people are asked to judge the career of the following
individual: "Steve is very shy and withdrawn, invariably helpful, but
with little interest in people or in the world of reality. A meek and
tidy soul, he has a need for order and structure, and a passion for detail." Research shows that people judge the probability that Steve is,
for example, a librarian by considering how similar Steve is to their
stereotype of a librarian. Serious errors of judgment result from ne-
glecting such factors as the percentage of librarians in the total population, a factor which Bayes' theorem requires us to introduce in the
form of a prior probability. Use of resemblance thinking also disposes
people to neglect other important features in the estimation of probabilities, such as sample size and regression. In a forthcoming book,
Richard Nisbett and Lee Ross describe numerous situations in which the
use of resemblance criteria leads people to make errors in attributing
behavior to spurious causes.[5]

The second piece of evidence for the pervasiveness of resemblance
thinking comes from anthropology. In his celebrated *The Golden Bough*,
Sir James Frazer cites a "Law of Similarity" as one of two principles
on which magic everywhere rests. This is the law that "like produces
like, effect resembling cause."[6] Frazer calls magic based on the law
Homeopathic Magic, and describes such applications as injuring or destroying the image of an enemy as a way of producing actual injury.
He provides numerous additional examples of homeopathic magic, in many
different cultures.

More recently, Richard Shweder has urged that we should understand the strange beliefs of other peoples as the result of applications of resemblance thinking.[7] For example, Zande beliefs about using fowl excrement to cure ringworm are viewed as part of a universal inclination to rely on resemblance instead of tests of correlation. Magical thinking is then viewed, not as an especially bizarre or primitive mode of thought, but as an application of a way of thinking all too natural to human beings.

The final support for the importance of the resemblance/correlation distinction comes from the historical work of Michel Foucault and Ian Hacking. Foucault states:

> Up to the end of the sixteenth century, resemblance played a constructive role in the knowledge of Western culture. It was resemblance that largely guided exegesis and the interpretation of texts; it was resemblance that organized the play of symbols, made possible knowledge of things visible and invisible, and controlled the art of representing them.[8]

According to Foucault, until the seventeenth century there was no distinction between what is *seen* and what is *read*. He quotes Paracelsus' assertion that God "has allowed nothing to remain without exterior and visible signs in the form of special marks."[9] Knowledge is then gained by reading the signs displayed in the world, and resemblance is the primary method for this. One application is the doctrine of signatures which guided medicine to such conclusions as that the lungs of the fox are an aid to the asthmatic, and that turmeric with its yellow color serves as a cure for jaundice.[10] Only in the seventeenth century, with the work of Descartes, Bacon and others, do we have the rise of correlation thinking which seems so fundamental to us today.

Foucault's views on the dominance of resemblance thinking are confirmed by Ian Hacking's work on the emergence of probability. Hacking marks the decade around 1660 as the birthtime of probability, in both its statistical aspect concerning frequencies and its epistemological aspect concerning degrees of belief.[11] Previously, "probability" indicated approval or acceptability by intelligent people; evidence was a matter of testimony and authority, not observation or correlation; resemblance thinking sufficed to interpret God's handiwork. The new dual concept of probability was thus part of the emergence of the method of correlation thinking.

A recent example of resemblance thinking is found in a *Time* magazine article on football.[12] It reports the work of a Berkeley

anthropologist who argues that the sexual symbolism of the game - team-
mates hugging and patting each other, the quarterback receiving the
ball from between the center's legs, talk of "scoring", skintight pants,
and so on - makes it clear that football is a homosexual ceremony,
serving to discharge the homoerotic impulses of players and fans. A
few similarities encourage an untested causal hypothesis.

Note however that not all reasoning involving similarity is resemblance thinking. The detection of similarities is a pervasive feature
of thought and only becomes illegitimate when the leap is made from
similarity to causal connection. Analogies and metaphors are very important in scientific and everyday reasoning but analogical inference
is not a form of resemblance thinking. In resemblance thinking we infer from the similarity of two things or events A and B that they are
causally related. Analogical inference involves similarity and causality, but in a more complicated fashion. We know that A and B are similar (analogous) to each other. We also know that A is causally related
to C; correlation thinking would have to be the basis for this knowledge. If C is similar to D, which is in the vicinity of B, we might
conclude by analogical inference that B might be causally related to D.
Our inference is not based on any similarity between B and D, but is
grounded in the known causal relation between A and C. To establish
more than the presumption of a causal relation between B and D would require further correlation. Analogy here functions primarily as a heuristic device, although it may also play a subsidiary role in the validation of the existence of a causal relation between phenomena.

Let us now turn to the question of pseudoscience and what differentiates it from science. Astrology is our paradigmatic example of a
pseudoscience. Because the use of resemblance thinking is so integral
to astrology, it is tempting to suggest that resemblance thinking is a
central feature of all pseudoscience. The temptation is strengthened
by the fact that the use of correlation thinking is so closely associated with the rise of modern science. However, the resemblance/corre-
lation distinction does not mesh so neatly with the pseudoscience/sci-
ence distinction.

First, not all pseudoscience uses resemblance thinking. Although
astrology and folk medicine revel in resemblances, it is also possible
to promulgate pseudoscience on the basis of spurious correlations. It
is a commonplace, especially among social scientists, that not all correlation indicates causation. There is a standard joke in introductory

logic books about the man who successively developed massive hangovers
from drinking scotch and water, gin and water, and vodka and water,
then prudently decided to give up drinking water. Falling barometers
do not cause storms. Thus mere attention to correlations is not sufficient to provide a scientific ground for a causal relation between
objects or events. Proponents of biorhythms delight in pointing to
such "confirmations" as that Elvis Presley died on a "triple low" day,
and that Mark Spitz won all his Olympic gold medals while experiencing
a "triple high." What I call correlation thinking involves much more
than attention to selected positive instances. It requires attention
also to negative instances as well as to possible alternative explanations of observed correlations.

The deans of modern pseudoscience, Immanuel Velikovsky and Erich
von Daniken, both use a very rough sort of correlation thinking in
support of their peculiar theories.[13] Velikovsky uses such evidence
as the coincidence among ancient myths to support his hypothesis that
Venus was ejected from Jupiter about 5,000 years ago and passed near
the earth before assuming its present orbit. Von Daniken also uses
mythological evidence in support of his hypothesis of ancient visits by
extraterrestrial beings. Both neglect alternative explanations and
suffer from other problems, such as the inconsistency of Velikovsky's
views with celestial mechanics, but the central objections to them do
not include the use of resemblance thinking. (There do however seem to
be a few instances, such as Velikovsky's view that the manna which supposedly nourished the ancient Hebrews during their years of wandering
in the desert was carbohydrates from the tail of Venus.) Thus it is
quite possible to be pseudoscientific without using resemblance thinking.

More problematically, we can look for uses of resemblance thinking
which are not pseudoscientific. Consider first the methods of the humanities, especially as they concern the interpretation of texts. Literary interpretation, art appreciation, and the history of philosophy
all in part involve the detection of similarities and the comparison of
symbols. But this study of resemblances is not pseudoscientific because it does not *purport* to be scientific; connections are found,
without making claims concerning causes and explanations. Although
there is occasional use of correlation methods, as in the computer analysis of texts to establish authorship, the humanities have an important function in providing *plausible* interpretations; truth is another

matter, one not to be reached by resemblance thinking.[14] To take an
example from the history of philosophy, the interpretation of Kant can
be a philosophically important enterprise even though we may have little hope of figuring out what Kant "really" meant.

Thus the humanities do not really provide any examples of the use
of resemblance thinking which is not pseudoscientific: as defined
above resemblance thinking is more than the recognition of similarities; it is the attribution of causality on the basis of similarities.
Interpretation in the humanities does not generally involve causal at-
tributions, so should not be counted as resemblance thinking. We
should say that the humanities are *non*-scientific rather than pseudo-
scientific, since their study of similarities has a function quite different from the attribution of causality.

One might suppose that biological classification uses resemblance
thinking. However, both the evolutionary and phenetic approaches to
taxonomy are founded on some sort of correlation thinking.[15] The former draws sustenance from the empirically supported theory of evolution, while the phenetic approach is highly quantitative.

Let us now take a brief look at Freudian psychoanalysis. Freud
and his followers would certainly claim that their theories are based
on correlation thinking applied to many clinical observations. But
controlled experiments are rare and much of Freudian theory is redolent
of resemblance thinking. The penis is elevated to such symbolic importance that female resentment of male domination can be brushed away
as penis envy. Compulsive neatness is attributed to problems of toi-
let training producing an anal retentive personality. A death instinct
is hypothesized to explain human self-destructiveness. Nisbett and
Ross, following Tversky, note Freud's great contribution of observing
that people frequently use resemblance thinking and developing the
method of free association to explore their thought processes; but
Freud himself seems to have fallen into resemblance thinking in pro-
pounding such doctrines as the ones mentioned. However, psychoanalysis at least attempts to use correlation thinking, and it might be argued that the psychological explanations which appear to be based on
resemblance thinking arise from noting the actual associations in
patients' minds; it would then be a correlational observation of peoples' use of resemblance thinking. Thus psychoanalysis, like textual
interpretation and biological classification, does not provide a

counterexample to the claim that all disciplines which use resemblance thinking are pseudoscientific.[16]

It thus appears that the use of resemblance thinking is a sufficient but not a necessary condition for pseudoscience. We can classify disciplines as follows:

1. Sciences: using correlation thinking.
2. Pseudosciences: (a) using resemblance thinking.
   (b) using inadequate correlation thinking.
3. Nonsciences: using neither resemblance nor correlation thinking, but dealing with non-causal matters such as textual interpretation or with normative issues, including ethics and aesthetics.

With this classification in mind, let us now return to the case of astrology.

In a recent paper, I proposed a new criterion of demarcation to answer the question of why astrology is a pseudoscience:

> A theory which purports to be scientific is pseudoscientific if and only if (1) it has been less progressive than alternative theories over a long period of time, and faces many unsolved problems, but (2) the community of practitioners makes little attempt to develop the theory towards solutions of the problems, shows no concern for attempts to evaluate the theory in relation to others, and is selective in considering confirmations and disconfirmations.[17]

This still seems to me to spotlight much of what differentiates astrology from genuine science. Astrology is stagnant, beset with problems, less promising than psychological theories of personality and behavior, and afflicted with uncritical proponents. Nevertheless, the criterion is too *soft* on astrology! For it implies that astrology only became a pseudoscience when the rise of psychology in the nineteenth century provided possible alternative explanations of personality. Such a condition makes sense in general, for I accept the arguments of Kuhn and Lakatos that there is no falsification in science without an alternative theory.[18] These arguments seem to imply that astrology could not be said to be false, let alone pseudoscientific, until alternative explanations were available. The mistake here is in taking astrology to be an established theory, one which passed a minimum *threshold* for acceptance and hence could only be dislodged by a better explanation. But if astrology is based on dubious resemblance thinking, providing

mere analogies rather than causal explanations, then there is no reason
to suppose that another theory is required to falsify it. Resemblance
thinking kept astrology from getting off the ground in the first place!
The use of resemblance thinking marked it as a pseudoscience well before it came to satisfy the criterion for pseudoscience stated above.

However, we have to leave open the possibility that astrology, the
study of cosmic influences on human personality and behavior, could *become* scientific. This does not appear to me very likely, but there are
attempts by Michel Gauquelin, Hans Eysenck and others to revitalize
astrology by statistical tests, and some correlations have been claimed.[19] We have here at least the *attempt* to use correlation thinking in
astrology, so that astrology may develop enough to deserve to be evaluated concerning its explanatory power and progressiveness in respect to
competing scientific theories. A rigorously revamped astrology based
on correlation thinking might then be judged to be a *protoscience*, a
discipline on its way to becoming a mature science, rather than a pseudoscience.

In sum, resemblance thinking is attribution of causal relations on
the basis of similarities. It is a pervasive aspect of human reasoning,
giving rise to much fallacious reasoning and contributing to such pseudosciences as astrology. The use of resemblance thinking is sufficient
to render a discipline pseudoscientific, but pseudosciences can also be
founded on incomplete or spurious correlation thinking. In the latter
case, we must use a more general demarcation criterion such as mine
quoted above. Obviously, the question of demarcation between science
and pseudoscience is immensely complicated. This should not be surprising in light of the complexity of science itself, and in light of
recent work in the theory of meaning which warns against the temptation
to seek necessary and sufficient conditions for the application of a
term.[20] Perhaps we should abandon the search for a demarcation criterion altogether.

But we certainly need not abandon in general the discussion of the
difference between science and pseudoscience, an abandonment which
would be especially regrettable in view of the *social* importance of
distinguishing science from the numerous exercises in irrationality
now prevalent in our culture.[21] To understand the nature of pseudo-
science, we must study stereotypical pseudosciences such as astrology,
comparing them with stereotypical sciences such as physics. The comparison reveals characteristics which we can at least take as marks of

pseudoscience, as criteria in a sense weaker than universal conditions.
Such characteristics include unprogressiveness, absence of critical
evaluation in the face of unsolved problems, and the matter I have discussed here: the use of resemblance rather than correlation thinking.[22]

FOOTNOTES

1 John Stuart Mill, *A System of Logic*, Longman, London, 1970, p. 501.

2 Lawrence E. Jerome, *Astrology Disproved*, Prometheus, Buffalo, 1977, p. 70.

3 See Michel Gauquelin, *The Scientific Basis of Astrology*, Stein and Day, Chicago, 1969.

4 Amos Tversky and Daniel Kahneman, "Judgment and Uncertainty: Heuristics and Biases," *Science*, 185, 1974, 1124-1131.

5 Richard Nisbett and Lee Ross, *Human Inference: Strategies and Shortcomings*, Prentice-Hall, Englewood Cliffs, in press.

6 Sir James Frazer, *The New Golden Bough*, ed. Theodor Gaster, Mentor, New York, 1964, p. 35.

7 Richard A. Shweder, "Likeness and Likelihood in Everyday Thought: Magical Thinking in Judgments about Personality," *Current Anthropology*, 18, 1977, 637-648.

8 Michel Foucault, *The Order of Things*, Vintage, New York, 1973, p. 17.

9 *Ibid.*, p. 26.

10 Cf. Mill, *op. cit.*, p. 502.

11 Ian Hacking, *The Emergence of Probability*, Cambridge University Press, Cambridge, 1975, pp. 11f.

12 "Football as Erotic Ritual," *Time*, November 13, 1978, p. 112.

13 I. Velikovsky, *Worlds in Collision*, Dell, New York, 1965. E. von Daniken, *Chariots of the Gods*, Putnam, New York, 1970. On Velikovsky see Carl Sagan, "An Analysis of *Worlds in Collision*," *The Humanist*, 37, 1977 (Nov./Dec.), 11-21.

14 On aesthetic plausibility see Dennis Dutton, "Plausibility and Aesthetic Interpretation," *Canadian Journal of Philosophy*, 7, 1977, 327-340.

15 For a summary of the two approaches, see Michael Ruse, *The Philosophy of Biology*, Hutchinson, London, 1973, chs. 7-8.

16 Of course, psychoanalysis would provide a counterexample only if it used resemblance thinking and was scientific. The scientific status of psychoanalysis is controversial: see Adolf Grünbaum, "How Scientific is Psychoanalysis?" in R. Stern, L. S. Horowitz, and J. Lyne, eds., *Science and Psychotherapy*, Haven Publishing, New York, 1977, pp. 219-254.

17 Paul Thagard, "Why Astrology is a Pseudoscience," in Peter D. As-
    quith and Ian Hacking, eds., *PSA 1978*, vol. 1, Philosophy of Sci-
    ence Association, East Lansing, Michigan, 1978, pp. 223-234.

18 Thomas Kuhn, *The Structure of Scientific Revolutions*, 2nd ed.,
    University of Chicago Press, Chicago, 1970. Imre Lakatos, "Falsifi-
    cation and the Methodology of Scientific Research Programmes," in
    I. Lakatos and A. Musgrave, eds., *Criticism and the Growth of Know-
    ledge*, Cambridge University Press, Cambridge, 1970, pp. 91-195.

    I have argued elsewhere that inference to the truth or falsity
    of scientific theories is "inference to the best explanation,"
    where a theory is accepted if it provides a better explanation of
    the evidence than competing theories; see Paul Thagard, "The Best
    Explanation: Criteria for Theory Choice," *Journal of Philosophy*,
    75, 1978, 76-92.

19 See Gauquelin, *op. cit.*, and J. Mayo, O. White, and H. J. Eysenck,
    "An Empirical Study of the Relation Between Astrological Factors
    and Personality," *Journal of Social Psychology*, 105, 1978, 229-236.

20 Hilary Putnam, *Mind, Language, and Reality*, Cambridge University
    Press, Cambridge, 1975, p. 271. Cf. Ludwig Wittgenstein, *Philoso-
    phical Investigations*, trans., G. E. M. Anscombe, Basil Blackwell,
    Oxford, 1968, paragraph 66. Abandonment of the view of meaning
    based on necessary and sufficient conditions does not *entail* that
    no general demarcation criterion can be found, since there might be
    such a criterion independent of the question of the meaning of the
    terms "science" and "pseudoscience," But it does undercut what
    seems to me to have been one of the main reasons for supposing that
    there is such a demarcation criterion.

21 Why is it socially important to distinguish science from pseudosci-
    ence? Roger Cooter, in a paper read to the Calgary Symposium,
    claimed that to use "pseudoscience" as a label is always to take a
    conservative stand. On the contrary: if pseudosciences such as
    astrology serve as opiates distracting people from a real under-
    standing of themselves and their society, where such an under-
    standing would contribute to desirable social change, then attack-
    ing pseudoscience can only be seen as politically progressive.
    Cooter seems to assume that talk of pseudoscience is reactionary be-
    cause it glorifies science, which he takes as part of capitalist
    ideology. But I would argue that there cannot be a critique of
    that ideology without an objective, i.e. scientific, understanding
    of the social system.

22 I am grateful to many participants of the Symposium, but especially
    Adolf Grünbaum and Marx Wartofsky, for helpful discussions. Mark
    Kaplan's suggestions about a threshold for inference to the best
    explanation also contributed.

THE ROLE OF PSYCHOLOGICAL EXPLANATIONS OF THE REJECTION OR ACCEPTANCE OF SCIENTIFIC THEORIES

In a public lecture at my University, the philosopher Michael Scriven challenged the credentials of psychoanalytic treatment. Immediately afterward, a senior psychoanalyst in the audience turned toward me to inquire whether Scriven's father or brother was an analyst. Evidently, the interlocutor deemed it unnecessary to come to grips with the lecturer's *arguments* for doubting the capability of available clinical evidence to sustain the claims of efficacy that had been made for Freudian psychotherapy.

Another colleague, concerned with the light that psychoanalytic principles might throw on some of the humanistic disciplines, concluded that the purported insights afforded by these principles are largely all-too-facile pseudo-explanations. Professional psychoanalysts present at university lectures in which he expounded this scepticism across the country usually responded rather patronizingly as follows: They offered diagnoses of the neurosis that had allegedly impelled the sceptical colleague to reject psychoanalytic theory after he presumably experienced ego-threat from it. Incidentally, the analysts in question repeatedly offered these dismissive psychological explanations with great confidence, undaunted by the fact that a Freudian diagnosis avowedly requires a considerable number of analytic sessions. Perhaps it is therefore not surprising that no two analysts offered the *same* diagnosis as to the sceptic's presumed neurotic affliction.

Far from being atypical, such psychologistic responses to criticism are, alas, rather representative, as illustrated by a comment on my forthcoming book *Is Psychoanalysis a Pseudo-Science?*, given to me by a practicing psychoanalyst. He pointed out that typical analyst readers will be looking not so much at my reasoning as at those of my psychological motivations purportedly discernible from my citation of derogatory assessments of analysis by others.

Indeed, I encountered the same dismissive psychologism as my sceptical colleague above when I recently examined the credentials of Freudian theory from a philosophy of science perspective in a lecture on a campus in Arizona. And the substantive issues posed in that Arizona encounter will now serve as my point of departure for dealing with

the central question of this paper, which is the following: Just what
ought to be the role, if any, of giving *psychological* explanations of
either the rejection or the acceptance of supposedly scientific theories
of man or nature? And similarly for the rejection or acceptance of
philosophical beliefs or religious doctrines.

The lecture I gave in Arizona had been largely sceptical as to the
rigor of the empirical validation of the Freudian corpus to date. But
I also defended psychoanalysis strenuously against the more damning
charge, leveled by Karl Popper, of being altogether untestable and
hence even unworthy of serious scientific consideration. Afterward, a
certified psychoanalyst who is a senior professor of psychiatry there
rose to rebut what she characterized as my "scurrilous attack" on
Freud's theory. Referring to Freud's own account of the reasons *and*
causes of opposition to psychoanalysis, she *overlooked* his explicit allowance for "those resistances to psycho-analysis that...are of the
kind which habitually arise against most scientific innovations of any
considerable importance."[1]

Instead of recognizing that Freud had thus made provision for the
existence of rational, evidential motivations for scepticism, she focused entirely on the fact that he had also identified irrational,
extraevidential or purely psychological inspirations of resistance to
his theory. According to Freud[2] the latter motivations

> are due to the fact that powerful human feelings are
> hurt by the subject-matter of the theory. Darwin's
> theory of descent met with the same fate, since it
> tore down the barrier that had been arrogantly set
> up between men and beasts. I drew attention to this
> analogy in an earlier paper (1917), in which I showed
> how the psycho-analytic view of the relation of the
> conscious ego to an overpowering unconscious was a
> severe blow to human self-love. I described this as
> the *psychological* blow to men's narcissism, and com-
> pared it with the *biological* blow delivered by the
> theory of descent and the earlier *cosmological* blow
> aimed at it by the discovery of Copernicus.

But as we saw, Freud himself had indeed made provision for the
existence of rational, evidential motivations for scepticism toward
psychoanalysis. Unfortunately at other times he simply evaded the
cognitive question of validation by pointing to the *de facto* growth of
acclaim for his theories and/or his mode of psychiatric treatment. For
example, he manifested just such an attitude when he commented on the
challenge to *demonstrate* the therapeutic efficacy of analytic treatment:
He was not only pessimistic regarding the feasibility of actually

demonstrating the efficacy of psychoanalytic treatment but gave a
strange twist to the cognitively unsolved question of efficacy by trans-
muting it into a sociological problem of resistance to new treatment
modes, which will solve itself with the passage of time. Thus, Freud[3]
wrote (in the unpolished English translation by J. Riviere):

> ...the social atmosphere and degree of cultivation of
> the patient's immediate surroundings have considerable
> influence upon the prospects of the treatment.
>
> This is a gloomy outlook for the efficacy of psy-
> choanalysis as a therapy, even if we may explain the
> overwhelming majority of our failures by taking into
> account these disturbing external factors! Friends of
> analysis have advised us to counterbalance a collection
> of failures by drawing up a statistical enumeration of
> our successes. I have not taken up this suggestion
> either. I brought forward the argument that statistics
> would be valueless if the units collated were not alike,
> and the cases which had been treated were in fact not
> equivalent in many respects. Further, the period of
> time that could be reviewed was too short for one to be
> able to judge of the permanence of the cures; and of
> many cases it would be impossible to give any account.
> They were persons who had kept both their illness and
> their treatment secret, and whose recovery in conse-
> quence had similarly to be kept secret. The strongest
> reason against it, however, lay in the recognition of
> the fact that in matters of therapy humanity is in the
> highest degree irrational, so that there is no prospect
> of influencing it by reasonable arguments. A novelty
> in therapeutics is either taken up with frenzied en-
> thusiasm, as for instance when Koch first published his
> results with tuberculin; or else it is regarded with
> abysmal distrust, as happened for instance with Jenner's
> vaccination, actually a heaven-sent blessing, but one
> which still has its implacable opponents. A very evi-
> dent prejudice against psycho-analysis made itself
> apparent. When one had cured a very difficult case
> one would hear: "That is no proof of anything; he
> would have got well of himself after all this time."
> And when a patient who had already gone through four
> cycles of depression and mania came to me in an inter-
> val after the melancholia and three weeks later again
> began to develop an attack of mania, all the members
> of the family, and also all the high medical authori-
> ties who were called in, were convinced that the fresh
> attack could be nothing but a consequence of the at-
> tempted analysis. Against prejudice one can do nothing
> as you can now see once more in the prejudices that
> each group of the nations at war has developed against
> the other. The most sensible thing to do is to wait
> and allow them to wear off with the passage of time.
> A day comes when the same people regard the same
> things in quite a different light from what they did
> before; why they thought differently before remains
> a dark secret.

        It is possible that the prejudice against the
analytic therapy has already begun to relax.

As a response to the challenge that he provide genuinely cogent
evidence for the therapeutic efficacy of analysis, Freud's statement
here is deplorably question-begging and evasive. Furthermore, previously withheld assent to a theory, finally given for avowedly unknown
reasons cannot be claimed to redound to the theory's *evidentially warranted* credibility. Even if such unexplained assent becomes widespread,
it cannot cogently be held to count in favor of the theory, any more
than ill-founded initial prejudice can validly count against it. Nonetheless even ill-founded prejudices for *or* against a theory may be evidentially *fruitful* as follows.

An emotional revulsion felt for a theory, and *alternatively* an attraction toward it, may *each* be conducive to uncovering *evidentially*
relevant information. Hostility toward the theory and/or toward its
advocates may inspire a successful and useful search for *objectively
contrary* evidence. But by the same token, the desire to embrace a
theory and/or loyalty to its exponents may issue in ferreting out bona
fide supporting evidence. Indeed, even the desire to *refute* a hypothesis may beget the unearthing of evidence objectively *favorable* to it
after all. As I recall from a lecture given by then Yale scientist
Cuyler Hammond, he had conducted his pioneering research on the effects
of cigarette smoking on humans in order to *refute* the conjecture that
smoking is harmful: Having been a chain smoker for years, he hoped for
evidence providing the comfortable assurance that he may continue to
enjoy his four packs per day with impunity. But *malgré lui*, he then
stumbled on stubborn evidence that drove him to the validation of a
link between (heavy) cigarette smoking and lung cancer! Again the
passionate desire to find evidence *favorable* to a given hypothesis in a
certain domain of occurrences *may* serve to uncover highly disconcerting
negative evidence.

So much for the possible heuristic benefits of emotional prejudice.

Now let us consider the merits of purely psychological appraisals
of the motives for the rejection of theories like psychoanalysis. Unfortunately, many psychoanalysts still need to be reminded that it is
illicit simply to *assume* the theory whose truth is first at issue, and
then to invoke this very theory as a basis for a psychologistic dismissal of *evidential* criticism of its validity. Freud did not claim
that psychoanalytic doctrine was handed to Moses on Mt. Sinai, and even

if he had, the rest of us may be forgiven for nonetheless asking concerning the evidence for it. One wonders how those psychoanalysts who *dismiss* evidential criticism in the stated psychologistic fashion react to the following analogously question-begging *theological* argument: Even insufficient or *prima facie contrary* evidence for the existence of God does *not* impugn His existence, because God is merely *testing us* by giving us insufficient evidence of His presence! In this vein, Martin Buber invoked the doctrine of "the eclipse of God" to reconcile the horrors of the holocaust with the goodness of God amid attributing any goodness in the world to divine beneficence.

As recently as 1948, the psychoanalyst Robert Fliess - who is not to be mistaken for Freud's own confrère Wilhelm Fliess - gave an ominously totalitarian twist to his advocacy of *psychologistically dismissing* the doctrinal fractionation of the Freudian movement into splinter groups and even into avowedly rival schools of thought.[4] He points out that even a highly trained analyst - no less than a run-of-the-mill analysand - may be neurotically resistant to the import of a Freudian tenet for his self-image. Especially when a future analyst is still undergoing his own training analysis, he may be prompted to question or even reject such a threatening Freudian tenet. Thus, the future analyst may

> vary the truth, instead of confirming it through improved introspection. In other words, his resistance may acquire the form of "dissension." He may, of course, do so in any phase of his education. He may, for instance, already working as an analyst, be confronted with equally unacceptable data by his patient, and be compelled by the consequent imminence of an empathic disturbance to gravitate, in the interest of his own psychic equilibrium, in the direction of an attenuated version of the theory of the unconscious. Or he may, in planning his training, unwittingly sidestep an as yet merely potential conflict of the same nature, by selecting a school of dissension, whose very existence derives from the same predicament in its founder. For it is here that the origin of dissension must be sought. The impact of psychic forces activated within an investigator in the course of his work may cause them to be reflected upon the collective object of his investigation and to direct his theoretical thinking. And if the personality of the individual thus imposed upon is a strong one, he is apt to break the ties of an inadequate education, and effect the foundation of a new school of psychological thought.
>
> There are, naturally, concurring motives for establishment of these "schools." The historical closeness to Freud, the consequent "personalization" of learning, and the demand to accept a

whole discipline practically from the hands of one
man of genius, are a challenge to anyone's independence of mind. It is this intellectual independence
which has indeed not infrequently led the dissenter-
to-be, before he yielded to the public demand for
the expurgation of psycho-analysis, to notable contribution to psycho-analysis in the sense in which
it is here discussed. Yet it has never done so
thereafter. For the denial of any one of the basic
and interdependent facts found by Freud cannot but
cause a defective crystallization of thought around
the hollow nucleus of negation...
    If, at a future time, the extravagant growth
of contemporary psychological teaching should be
pruned back to the live stem of observation and
theory of the first, the Freudian, period of its
existence, psycho-analysis will regain its original independence of the preconceptions of the
general as well as the learned public. It will
then acquire the status of a scientific discipline
comparable to others. Scholars from various fields
will be given the opportunity to become competent
in it, and, putting an old Jesuitic practice to
secular purpose, will school themselves in both
disciplines, psycho-analysis and their own. And
an elaborate post-graduate education will be preventive of wasteful effort by including in its
requirements that dissension be generally subjected to clearance in an analysis supplementary
to the training analysis of the dissenter.[5]

Invoking the Freudian doctrine of neurotic resistance to the recognition of buried inner conflicts, R. Fliess assumes in egregiously
question-begging fashion that Freud always knows best, no matter what
future evidence might be offered critically by an analyst at any stage
of his professional development. Clearly, Fliess's automatic and *advance* discounting of *all* scepticism, however documented, as being solely inspired by neurotic resistance turns the validation of psychoanalysis into a logically circular *self*-validation. And it apparently did
not occur to him - as Freud himself was driven to appreciate à propos
of his seduction aetiology of hysteria and of his sexual aetiology of
obsessional neurosis - that this vicious circle can be broken as follows: One can utilize the probative import, be it favorable or ad-
verse, of evidence from the kinds of *extra*-clinical events that simply
do not provide scope for the contaminating intrusion of resistance.

    Alas, Fliess's rationale for his cowardly euphemistic "clearance"
of dissenting intellectual independence is highly reminiscent methodologically of the candidly labeled "sacrifice of the intellect" enjoined
by Ignatius of Loyola as the highest grade of obedience (in his 1553
Letter on Obedience to the Jesuits of Coimbra). And Fliess's proposed

management of dissent is a cognate of the penchant of Soviet security
organs to view political dissent as a psychiatric problem. On the
other hand, no less authoritative a psychoanalytic spokesman than
Ernest Jones repudiated the paternalistic authoritarianism espoused so
patronizingly by Fliess. Speaking of the conclusions reached by Freud
on the basis of his investigations, Jones declared: "...it is plain
that we should be forsaking the sphere of science for that of theology
were we to regard these conclusions...as being sacrosanct and eternal."[6]

In the same vein, the well-known analyst Edward Glover, deploring
precisely the dismissive appeal to neurotic "resistance" employed by
Fliess, writes:[7]

> It is scarcely to be expected that a student who
> has spent some years under the artificial and
> sometimes hothouse conditions of a training ana-
> lysis and whose professional career depends on
> overcoming "resistance" to the satisfaction of
> his training analyst, can be in a favourable
> position to defend his scientific integrity
> against his analyst's theories and practice.
> And the longer he remains in training analysis,
> the less likely he is to do so. For according
> to his analyst the candidate's objections to
> interpretations rate as "resistances." In
> short there is a tendency inherent in the train-
> ing situation to perpetuate error. Such a state
> of affairs clearly calls for the application of
> special safeguards.

Yet it is unclear not only just how the latter safeguards are to function but also what Glover expects from their employment: He implicitly
provides ammunition for Fliess's dismissive stance by speaking of "our
proven aetiological systems,"[8] despite having acknowledged that the
analyst's interpretations of patient responses cannot be reliably check-
ed and thus constitute "the Achilles heel" of psychoanalytical investigation.[9]

If one accepts Ernest Jones' own account[10] of the intellectual and
personal rift between Freud and Adler, Freud's conduct toward Adler did
not violate Jones' aforecited lofty methodological injunction. Speaking of the fact that Adler left the Vienna Psychoanalytic Society and
formed his own splinter "Society for Free Psychoanalysis," Jones de-
clares[11] that "the freedom of science...is certainly a worthy cause."
But Jones adds:[12]

> The only issue was whether it was profitable to
> hold discussions in common when there was no
> agreement on the basic principles of the subject-
> matter; a flat-earther can hardly claim the *right*

to be a member of the Royal Geographical Society
and take up all its time in airing his opinions.
Adler had drawn the correct inference by resign-
ing. To accuse Freud of despotism and intoler-
ance for what had happened has too obvious a
motive behind it to be taken seriously.

Yet further documentation from the minutes of the Vienna Psychoanalytic Society shows the following:[13] While Freud did indeed offer a
relevantly argued rebuttal to Adler's critique of his theory, Freud
also engaged in a *question-begging* advance psychologistic dismissal of
any assent to Adler's doctrine by others. Freud did so when he pre-
dicted that this dissident doctrine "will make a deep impression and
will, at first, do great harm to psycho-analysis." Said he:[14] "it
offers general psychology. It will, therefore, make use of the latent
resistances that are still alive in every psychoanalyst, in order to
make its influence felt."

According to Colby's account,[15] not only Freud's conduct but that
of the membership of the Vienna Psychoanalytic Society was without blem-
ish at least to the following extent: Far from expelling Adler from
membership in the Society, as alleged by the biographers P. Bottome and
F. Wittels, a majority of that Society voted to express regret over
Adler's departure from it. But even the Freudian partisan Jones acknowledges[16] that, at a special plenary session of the Society on October
11, 1911 (Meeting #146) at which Freud announced the resignation of
Adler and of three others, the following was voted by a majority of
eleven to five: No one is to belong to both that Society and to the
dissident one founded by Adler. And as Jones notes,[17] this affirmation
of "a strong desire for a clean break" then immediately issued in the
resignation of the remaining six pro-Adlerians from the original Society.

Having cited the version of the Freud-Adler rift furnished by the
pro-Freudian authors Jones and Colby, it behooves us to summarize the
salient points from the documented account just furnished by the historian of science Janet Terner and the psychiatrist W. L. Pew, who are
Adlerian partisans.[18]

In 1910, Alfred Adler became president of the Vienna Psychoanalytic Society as well as coeditor (with Wilhelm Stekel) under Freud of a
new journal for psychoanalysis. And early in 1911, at Freud's invita-
tion, Adler presented his critique of Freud's sexual theory in a series of three lectures. After a mass denunciation of Adler by the Freudians that was reportedly "almost unequaled in its ferocity,"[19] Adler

resigned his editorship as well as from the Vienna Psychoanalytic Society in the summer of 1911. And members of that Society who sided with Adler began informal gatherings with him at the Café Central. But at the next meeting of the Freudian group that fall

> Hanns Sachs read the indictment against Adler, and
> moved that it was incompatible to belong to both
> groups...the motion was carried and [the] six
> Adlerians rose, left, and went to the Café Central
> where [they] celebrated with Adler.[20]

Moreover

> Freud proscribed the quoting of Adler in any paper
> published by a Freudian (although Freud himself
> polemically railed against Adler whenever he chose).
> Indeed, one can examine the vast literature pro-
> duced by the Freudians over the decades and rarely
> find Adler mentioned. Yet strangely, Adler's con-
> cepts often appear in thinly disguised form and in
> some cases almost verbatim renditions.[21]

Still worse,

> Freud never forgave Adler's dissidence, and in the
> years that followed he wielded his eloquent pen to
> encourage his loyal followers to deny or discredit
> Adler's discoveries - to keep Adler forever in the
> shadow of his own greatness. Freud knew the power
> of legend, and when he wrote his history of the
> psychoanalytic movement, which was reprinted in
> the United States [footnote omitted] in 1916, he
> presented his own prejudiced version of the split
> with Adler, scurrilously attacked him, and judged
> his ideas as "radically false." In this way, he
> got his account of these events into the history
> books long before it seemed important to anyone
> else [footnote omitted]. This affected Adler's
> image in America both immediately and in the
> long run.[22]

As an example of the hostile reception accorded to Adler by the power brokers of psychoanalysis in the United States, Terner and Pew[23] cite the following 1916 indictment, which presages Robert Fliess's aforecited *psychologistic* dismissal of all doctrinal apostasy by one-time Freudians:

> Dr. James J. Putnam, a venerable Boston Brahmin
> and champion of Freud, wrote: "A great longing
> has been felt by many conscientious students of
> human nature to find some way of escape from ac-
> cepting Freud's conclusions....To such persons
> Adler's mode of explanation is only too attrac-
> tive. In plain terms, it offers a weapon with
> which Freud may be conveniently struck down by
> those...so minded [footnote omitted].

To illustrate just how "the 'hot' battle in Vienna was transplanted to America as a cold shoulder," Terner and Pew write:[24]

The impact of Freud's curse on Adler was
notable, and when Adler's *Study of Organ Infer-
iority* and *The Neurotic Constitution* appeared
the following year, the reviews in the psychiatric literature were few and terse, with one
notable exception....
But the die was essentially cast for Adler's
place within American psychiatry....To mention
Adler or openly advocate his ideas was a risky
stance against the mainstream of the profession.

No wonder, therefore, that when Rudolf Dreikurs - who was to become a
vigorous opponent of the Freudian *monopoly* in American psychiatry - arrived in New York in 1937, "He was firmly warned not to declare himself
an Adlerian,"[25] since *Freudian* psychoanalysis had become the mainstream
in psychiatry and indeed had virtually achieved hegemony over the entire spectrum of the mental health professions:[26]

To be part of the mainstream was considered vital
to most practicing psychiatrists. Its leaders
were the power brokers who held the key to hospital appointments, professorships, and publishing
opportunities - in sum - to recognition and success.

And by then, the psychologistic dismissal of opposition to Freud's ideas
by his disciples had become inveterate:[27]

As psychoanalysts became preeminent, they grew
less tolerant of opposing points of view.
Psychoanalysis was loudly acclaimed as a
scientifically proven body of theory and practice, and those who actively challenged or opposed it within the professions were discredited as the simple unwashed - that is, the unanalyzed and therefore unknowing and superficial.
Even eclecticism became ensnared in the defense
of psychoanalysis. Since psychoanalysis was regarded as an established truth, it therefore became "uneclectic" and dogmatic *not* to acknowledge
its basic truths. In other words, if these
truths were acknowledged and the catechism of
Freudian phrases was repeated, one could acquire
the luster of revered "objectivity."

In any case, purely psychological, *extra*-evidential explanations
of resistance to psychoanalysis *would* become relevant to understanding
its rejection, *if* the presupposition of the following question *were* in
fact true, which it is *not*: In the face of the doubting Thomases' explicit admission that there is indeed strong supporting evidence for
the theory, why do they nevertheless still deny its credibility? By
the same token, if Mr. X rejects atheism in favor of theism amid saying
himself that the pertinent evidence does favor atheism, then we can try
to understand his rejection in purely extra-evidential psychological

terms. But before inquiring into purely psychological motivations for
the rejection of either Freudism or any other "ism," the validity of
the given theory must be adjudicated on the basis of the balance of the
weight of the evidence. Until and unless this is done, the invocation
of purely psychological, extra-evidential explanations for *either* the
rejection *or* the acceptance of the theory runs the risk of begging the
question of its validity, if only because *either attitude may well be
prompted by relevant evidence*.

In saying this, I allow, of course, that the *available* evidence
may warrant the rejection of an actually true theory, or alternatively,
the acceptance of an actually false one. That is why there are what
physicians call respectively "false negatives" and "false positives,"
errors to which statisticians refer as "Type I" and "Type II" errors
respectively. But for the limited purpose of our present inquiry into
understanding the rejection or acceptance of a theory vis-à-vis its
validity in the light of the evidence, we can simple-mindedly lump together true beliefs with evidentially warranted ones on the one hand,
and false ones with evidentially unwarranted ones on the other. Then I
can say that psychological causation *as such* does not discriminate between valid beliefs and invalid ones. As I wrote elsewhere:[28]

> ...both true beliefs and false beliefs have
> psychological causes. The difference between a
> true or warranted belief [on the one hand] and a
> false or unwarranted one [on the other] must
> therefore be sought in the particular *character*
> of the psychological causal factors which issued
> in the entertaining of the belief; a *warrantedly
> held belief, which has the presumption of being
> true, is one to which a person gave assent in
> response to awareness of supporting evidence.*
> Assent in the face of awareness of a *lack* of
> supporting evidence is irrational, although
> there are indeed psychological causes in such
> cases [as well] for giving assent. Thus, one
> person may be prompted to give assent to a cer-
> tain belief solely because this belief is wish-
> fulfilling for him, while another may accept the
> same conclusion in response to his recognition
> of the existence of strong supporting evidence.
> And the belief *may* well be true.

More generally, as I wrote in the same essay,[29]

> the causal generation of a belief does not, of
> itself, detract in the least from its truth.
> My belief that I address a class at certain
> times derives from the fact that the presence
> of students in their seats is causally inducing
> certain images on the retinas of my eyes at

those times, and that these images, in turn, then
cause me to infer that corresponding people are
actually present before me. The reason why I do
not suppose that I am witnessing a performance of
Aïda at those times is that the images which Aïda
Radames, and Amneris would produce are not then in
my visual field. The causal generation of a belief in no way detracts from its veridicality. In
fact, if a given belief were not produced in us by
definite causes, we should have no reason to accept
that belief as a correct description of the world,
rather than some other belief arbitrarily selected.
Far from making knowledge either adventitious or
impossible, the deterministic theory about the origin of our beliefs alone provides the basis for
thinking that our judgments of the world are or may
be true. Knowing and judging are indeed causal processes in which the facts we judge are determining
elements along with the cerebral mechanism employed
in their interpretation.

On this causal conception of the generation of (perceptual) beliefs, there are typically at least some *extra*-ideational causes of the
occurrence of the following kind of awareness-state: A mental state
that is at once the evidential source (or "reason") and the psychological cause of entertaining (or espousing) a belief. Thus, the *"initial"*
causal promptings of our beliefs concerning the external world are
typically extra-ideational events rather than mental apprehensions of
evidential reasons. Hence I must reject the upshot of the following
account offered by C. S. Lewis:[30]

All beliefs have causes but a distinction must be
drawn between (1) ordinary causes and (2) a special
kind of cause called 'a reason'. Causes are mindless events which can produce other results than
belief....A belief which can be accounted for entirely in terms of causes is worthless. This principle must not be abandoned when we consider the
beliefs which are the basis of others. Our knowledge depends on our certainty about axioms and
inferences. If these are the result of causes,
then there is no possibility of knowledge. Either
we can know nothing *or* thought has reasons only,
and no causes.
    ...All attempts to treat thought as a natural
event involve the fallacy of excluding the thought
of the man making the attempt.
    It is admitted that the mind is affected by
physical events...But thought has no father but
thought. It is conditioned, yes, not caused....
    The same argument applies to our values,
which are affected by social factors, but if they
are caused by them we cannot know that they are
right.

On this basis, Lewis[31] feels entitled to give an affirmative answer to his question "Does 'I know' involve that God exists?". He invokes God as the very source rather than merely as the epistemological under-writer of our knowledge. Whereas Descartes' epistemological specter was the evil deceiving genius, Lewis' corresponding nightmare is con-stituted by "mindless events": Allegedly such events cannot causally induce bona fide knowledge states in us because "thought has reasons only, and no causes." How then can he hope to show that the inclusion of initially "mindless" causation does not significantly contribute to the warrant for the following well-taken assertion by him?: "Suppose I think, after doing my accounts, that I have a large balance at the bank. And suppose you want to find out whether this belief of mine is 'wish-ful thinking'. You can never come to any conclusion by examining my psychological condition."[32] But if so, is the initially "mindless" causation by corresponding actual bank assets held in his name not con-siderably more justificatory here than the much more inscrutable "Super-natural" of which he speaks?

But there is full agreement with C. S. Lewis[33] that, regardless of the motivation for holding a certain belief or disbelief in a given case, its *validity* must be assessed on the basis of the pertinent evi-dence, not by reference to the psychological motivation for entertain-ing it. It would be altogether fallacious, for example, to cast asper-sions on *disbelief* in personal immortality on the purported psychologi-cal ground that this disbelief is caused by the "death instinct" - Freud's "thanatos." Such psychologistic dismissal of the rejection of personal immortality might invoke man's putative craving for death as a release from the sorrows of life. But even if all men could be shown to harbor such a "death instinct," this psychological fact could not itself preclude that (a) there might *also* be strong objective evidence for *disbelieving* in the *post mortem* existence of the self, and (b) even psychologically, actual disbelief in an after-life is explained by awareness of this evidence rather than by the putative death instinct.

Similarly, the proponent of psychoanalysis in Arizona patently *begged the question* of its validity, when she sought to *dismiss evi-dential criticism* on the basis of the following purported psychological explanation: Those who question the evidential credentials of Freudian theory do so because it poses a so-called "narcissistic" threat to man's ego by asserting the sovereign dominance of unconscious forces, at least over unanalyzed people. As an intended premise of an *argument*

against disbelief in Freudian theory, this *dismissive psychological explanation* is thus clearly *irrelevant*. But furthermore, even as a purely causal account of opposition to psychoanalysis, this psychological
explanation fails *empirically*, since it does *not simultaneously* accommodate the impressive widespread *acceptance* of Freudianism during the
past quarter century, at least in the United States: After all, not
only in influential literary circles but even in sizable segments of the
educated lay public, psychoanalysis has become a cultural idol by com-
manding a kind of quasi-religious veneration.[34]

For example, in November of 1977, the journalist Jack Anderson published a report of an hour's interview of President Carter entitled
"What is Jimmy Carter Really Like?"[35] Anderson relates that after the
interview, he submitted the transcript to the psychoanalyst Saretsky.
And Anderson reports the conclusions reached by Dr. Saretsky after several days of study. One such product of Saretsky's expertise is that
Carter is a man who believes in God *and isn't afraid to say so*. Presumably, if Anderson felt less deferent toward psychoanalysts, he would
hardly have deemed it remarkable that a president of the United States
who does believe in God *isn't* afraid to say so. Oddly enough there was
no mention of Carter's earlier *Playboy* interview. Jack Anderson's reci-
tation of a series of other *at best* trite and altogether safe comments
from his analytic consultant typifies the halo that is often uncritical-
ly bestowed on psychoanalysis in our culture. Hence I am prompted to
ask: If the "narcissistic" threat to man's ego is held to account ade-
quately for the vigor of such opposition to psychoanalysis as is encountered in some quarters, then how could Freud's ideas have so tri-
umphantly swept aside these alleged defensive reactions in the culture
at large, even among many of the unanalyzed? Thus it would seem that
the all too facile invocation of ego-threat to explain *opposition* to
psychoanalysis also boomerangs *empirically* by running afoul of the
rather prevalent espousal of Freudianism.

In any case, the stated fallacy of *psychologistically* discrediting
the *rejection* of Freudianism is on a par with *each* of the following two
similarly dismissive gambits: (i) Endeavoring to undermine the espousal of atheism by claiming that Madeline Murray-O'Hare and other atheists simply *hate* the idea of a personal God, and (ii) seeking to discredit the advocacy of theism by merely pointing out that its adherents
derive much emotional solace from their belief in a cosmic protective
father figure. Freud[36] commendably recognized in his book *The Future*

of an Illusion that theism cannot validly be discredited by the latter dismissive gambit, and Erich Fromm usefully reiterated this point[37] in his book *Psychoanalysis and Religion*. We see that the attempt to undermine the theist's *acceptance* of belief in God purely *psychologistically* because it is wish-fulfilling is just as ill-conceived logically as the correspondingly dismissive endeavor to cast psychologistic aspersions on the atheist's *rejection* of that belief. Surely there are some beliefs that are both true and wish-fulfilling for at least a good many people, no less than there are other true beliefs that are wish-contravening.

   Mutatis mutandis, efforts to discredit the *espousal* of Freudianism by facile psychologistic devices are no less unsound logically than appealing to the ego-threat hypothesis as a basis for impugning intellectual *opposition* to psychoanalysis. I am concerned to stress this *equi*-fallaciousness. But so far, I have been at pains one-sidedly to indict the use of mere psychologism for dismissing the *rejection* of Freudian theory. Hence I now wish to call attention to the like unsoundness of the following peremptory psychologistic dismissal of the *acceptance* of psychoanalysis: Freud's bizarre adult psychoanalytic beliefs can be discounted without regard for his supportive arguments because - as we know from his biography - he experienced the traumatic childhood event of having entered his parents' bedroom while they were locked in a sexual embrace and of having been irately ordered out by his father.[38] By the same token, it will not do at all to try to impugn the therapeutic efficacy of Freudian treatment by offering the following *motivational* explanation of why anyone chooses to become a psychoanalyst: Analysts are driven by a kind of generalized voyeurism, coupled with morbid curiosity or are just socially-sanctioned peeping Toms. Unfortunately, none other than Thomas Szasz saw fit to stoop to the nadir of the psychologistic dismissal of Freud's theories: In the manner of gutter journalism, he depicts these theories as mere shams that purportedly owe their principal inspiration to Freud's desire to be "The Jewish Avenger" vis-à-vis the gentile world.[39]

   Recently the philosopher Frank Cioffi[40] has offered a *differently-argued* yet motivational critique of Freud's championship of psychoanalysis. What makes Cioffi's critique interesting is that though motivational, it is *not peremptorily* psychologistic. Thus Cioffi tried to *show* that while Freud seemingly went through the motions of being engaged in an explanatory inquiry, the Viennese doctor's ostensible

arguments become *intelligible* only after the following is recognized:
His hypotheses are prompted not by a concern with logically relevant
evidence, but by his idiosyncratically intense personal preoccupation,
presumably with sexuality.

But does Cioffi give cogent grounds for his claim that Freud's
reasoning is unintelligible, unless we *exclude* concern with pertinent
evidence from Freud's actual motivation for espousing psychoanalysis?
I shall briefly scrutinize just a couple of the considerations adduced
by Cioffi for this claim in order to show why I deem it to be ill-
founded. Thus I shall reject Cioffi's grounds for resorting to a purely
psychological explanation of Freud's own advocacy of psychoanalysis.
Cioffi sees himself as having resorted to explaining Freud's rationale
in terms of *extra*-evidential motives only *after* having ruled out logically pertinent evidential promptings. But I shall now illustrate how
he mishandled his examination of Freud's reasoning and was thereby
driven to the gratuitous or mistaken conclusion that concern with pertinent evidence had played no essential role in Freud's rationale for
espousing psychoanalysis.

Cioffi[41] gives the following lucid statement of his thesis:

> Freud behaves neither like someone who is address-
> ing himself to the problem of the causes and nature of
> the neuroses but bungles the job from incompetence or
> lack of methodological sophistication, nor like some-
> one who is stymied by the intrinsic difficulties of
> the problem, but rather like someone who, while going
> through the motions of engaging in an explanatory
> enquiry, reveals in an enormous variety of ways that
> he has other ends in view. (Fine shades of misbe-
> haviour.)
>
> One often comes across people whose preoccupa-
> tions with a putatively explanatory factor is osten-
> sibly derived from their interest in its pathogenic
> potentialities, but is really intrinsic. The major-
> ity of those who speculate as to whether slums, or
> the decline in church-going are causes of delinquency
> are not really interested in delinquency, but are
> interested in slums and the decline in church-going.
> What is noteworthy in Freud is the way in which the
> prestige of aetiological and prophylactic enquiries
> is exploited in the interest of an idiosyncratic
> preoccupation (or, perhaps I should say, an idio-
> syncratically intense preoccupation).

Having adduced Freud's psychobiography of Leonardo da Vinci in support of this thesis, Cioffi[42] writes as follows:

> As another instance of the ease with which the
> variety of mechanisms at Freud's disposal enables
> him to press into the service of the thesis of

infantile pathogenicity, whatever parental circumstances the childhood history of his subject happens to provide, consider his account of how inevitable it was, given the character of Dostoyev-
sky's father, that he [the son] should have come
to possess an over-strict super-ego: 'If the
father was hard, violent and cruel, the super-ego
takes over these attributes from him, and in the
relations between the ego and it, the passivity
which was supposed to have been repressed is re-
established. The super-ego has become sadistic,
and the ego becomes masochistic, that is to say,
at bottom passive in a feminine way. A great need
for punishment develops in the ego, which in part
offers itself as a victim to fate, and in part finds
satisfaction in ill-treatment by the super-ego (that
is, the sense of guilt).'

This is not at all implausible. But neither is this:

'The unduly lenient and indulgent father fosters the
development of an over-strict super-ego because, in
the face of the love which is showered on it, the
child has no other way of disposing of its aggres-
siveness than to turn it inwards. In neglected
children who grow up without any love the tension
between ego and super-ego is lacking, their aggres-
sions can be directed externally...a strict conscience arises from the co-operation of two factors
in the environment: the deprivation of instinctual
gratification which evokes the child's aggressive-
ness, and the love it receives which turns this ag-
gressiveness inwards, where it is taken over by the
super-ego.'

As *I* see it, the conjunction of these two quotations from Freud is tan-
tamount to the following assertion of *causal sufficiency*: If a child
has a father who is either "hard, violent and cruel" or "unduly lenient
and indulgent," then it develops a hyper-strict super-ego. Yet immediately after having given us the two Freud quotations in question,
Cioffi draws the following astonishing inference:[43]

That is, if a child develops a sadistic super-ego,
either he had a harsh and punitive father or he had
not. But this is just what we might expect to find
if there were no relation between his father's character and the harshness of his super-ego.

But Cioffi employs a gross logical sleight-of-hand when he trivi-
alizes Freud's stated *bicausal* claim into one that does *not* affirm any
causal relevance of the father's character to the child's super-ego.
For Cioffi replaced the non-trivial antecedent that the child's father
was either harsh or *unduly lenient* - which is a *non*-exhaustive disjunction - by the utterly trivial consequent that either the child had a
harsh father or he had not. And the latter disjunction is, of course,

trivial because it is exhaustive besides being mutually exclusive. In
other words, Cioffi has speciously replaced the paternal property of
undue leniency, which Freud had invoked causally, by the logically much
weaker property of merely being *non*-harsh. And amid committing this
sleight-of-hand, he likewise overlooked the following: Freud's bicausal assertion had affirmed the causal *sufficiency* of the specified
paternal traits for hyper-strict childhood super-ego development, but
not their being causally *necessary* for such super-ego development. For
Cioffi gratuitously depicts Freud as having legitimated a retrodictive
causal inference *from* the child's super-ego structure *to* that of the
father, whereas the Freudian quotations adduced by Cioffi countenance
only an inference in the opposite direction.

By parity with Cioffi's fallacious reasoning one could deduce the
absurdity that getting shot is causally irrelevant to being killed from
the following sound bicausal assertion: If a person is either shot in
a vital organ or massively poisoned by a fast-acting toxin *without* being shot in a vital organ, then he will be killed in all likelihood.
By parity of reasoning with Cioffi's argument, this would become: If a
person is killed, then he was either shot or he wasn't, which is just
what we would expect, if there were no relation between getting shot
and getting killed.

As a second brief illustration of Cioffi's attempt to document
Freud's alleged indifference to the actual evidence, let me cite another argument by him. Cioffi writes:[44]

> It might seem that there can be no question of the
> genuinely empirical-historical character of those
> clinical reconstructions which incorporate refer-
> ences to the external circumstances of the patient's
> infantile life, such as that he had been threatened
> with castration or been seduced, or seen his parents
> engaged in intercourse. These at least are straight-
> forwardly testable, and their accuracy would there-
> fore afford evidence of the validity of psychoanaly-
> tic method; for if the investigation into the infan-
> tile history of the patient revealed that he had no
> opportunity of witnessing intercourse between his
> parents (the primal scene), or that he [had] not
> been sexually abused, or not threatened with cas-
> tration, this would cast doubt on the validity of
> the interpretative principles employed and on the
> dependability of the anamnesis which endorsed them.
>   But Freud occasionally manifests a peculiar
> attitude towards independent investigation of his
> reconstruction of the patient's infantile years.
> In 'From the history of an infantile neurosis'
> [the case of the "Wolf Man"] he writes: 'It may

> be tempting to take the easy course of filling up
> the gaps in a patient's memory by making inquiries
> from the older members of the family: but I cannot
> advise too strongly against such a technique...One
> invariably regrets having made oneself dependent
> on such information. At the same time confidence
> in the analysis is shaken and a court of appeal is
> set up over it. Whatever can be remembered at all
> will anyhow come to light in the course of further
> analysis.'

But does Freud here reject the use of any and all independent external
evidence to test the historical veracity of his clinical reconstructions of a patient's infancy *after* the completion of the psychoanalysis?
It would seem not. What Freud does renounce here is reliance on inquiries directed by the analyst at older members of the patient's family to fill up the gaps in the patient's memory as a procedural technique for making progress in the analysis. This renunciation does not
preclude his willingness to test his clinical reconstructions by means
of independent external evidence once the analysis has been completed.
Indeed, he displayed this kind of willingness in his retrospective external evaluation of his clinical findings in his "Rat Man" case. For
in this case, he was prompted by just such independent historical evidence to abandon his prior hypothesis as to the *specifics* of the sexual
aetiology of adult obsessional neurosis. Corresponding remarks apply
to his abandonment of his seduction aetiology of hysteria, as we know
from his Letter 69 to Wilhelm Fliess, dated 21.9.1897.

Indeed this reading of Freud is strengthened upon taking account
of the sentences that Cioffi *omitted tendentiously* from his quotation.
The latter is taken from a footnote that Freud appended [cf. Collected
Papers, tr. A. and J. Strachey, 1959, Basic Books, New York, vol. 3,
pp. 481-2, fn. 2] to the statement that in the Wolf Man's later years,
this patient "was told many stories about his childhood." Freud's
appended footnote *begins* with the sentence "Information of this kind
may, as a rule, be employed as absolutely authentic material," a sentence that immediately precedes the one with which Cioffi begins his
quotation! Moreover, the sentence whose omission from *within* his quotation Cioffi does indicate reads as follows: "Any stories that may be
told by relatives in reply to inquiries and requests [from the analyst]
are at the mercy of every critical misgiving that can come into play."
The latter omission supplies Freud's *epistemological reason* for *not*
succumbing to "the easy course of filling up the gaps in a patient's
memory by making inquiries from the older members of his family":

Whereas, "as a rule," stories that the patient's relatives told the patient *spontaneously* in his later years are "absolutely authentic," responses by relatives to pointed inquiries from the analyst may well be quite contaminated by misgivings.

Hence, I believe that Cioffi's motivational critique of Freud misfires by being ill-founded. But this is *not* to say that *psychological* explanations of the acceptance of psychoanalysis are *always* misdirected.

For suppose it *were* now agreed that there is in fact a serious dearth of objective evidential support for psychoanalysis. And assume further that some educated people who do acknowledge this sparsity of validating evidence nonetheless vigorously espouse Freud's theory. Of course in that case we cannot attribute their assent to their awareness of cogent supportive findings, or to their belief that they possess such favorable evidence. Hence it then becomes well-nigh imperative that we ask: In the face of the avowed scarcity of such validating findings, why does anybody nonetheless embrace the theory? And plainly, under these conditions, it is hardly question-begging or misdirected to ask the following advisedly loaded question: What non-evidential psychological motivations prompt otherwise rational people to believe the fable brilliantly concocted by the genius whom Nabokov called "the Viennese witch doctor"?

Quite naturally, therefore, sceptics who do decry Freudian theory as a fable or pernicious myth have asked just this question. And some of them have suggested some possible answers. Let me conclude by concisely outlining four of their proposed explanations, although I do not venture to guess at their relative potential importance.

(a) Psychoanalysis as a general theory of man and even liberating gospel became the secular equivalent or alternative of traditional religion for many intellectuals who had abandoned theism. Being such a substitute religion, it commanded the fierce loyalty often displayed by those who are converted to a new religion during adulthood.

(b) As Salter[45] has pointed out:

> Writers as a group are probably among the most neurotic in the population. When word of the new panacea for their troubles with old wives or new books drifted through they went for the treatment in a big way. Many of them were encouraged to write about their analysis as a means of paying for the expensive treatment. A rash of novels, plays, and short stories resulted. Some even wrote lengthy magazine articles embedded with a remarkable variety of rather too-candid clinical detail.

> The net result was a public relations cam-
> paign that millions of dollars could not have
> duplicated. Once analysis became fashionable
> among the writers, it was a brief step before
> their more impressionable readers were fretting
> impatiently in the analysts' busy waiting rooms.

(c) The favorable retrospective valuation of the personal analyses
undergone by many professional people is often assured by the expense,
time and emotional pain of an analysis. "There is abundant evidence
that where we have made a sacrifice to obtain some object, we come to
value it: we cannot afford to admit to ourselves that we have made the
sacrifice in vain."[46] It is small wonder that relatively few people
are willing to admit to themselves that they *may* have made a costly
mistake in the sense that their actual therapeutic gains may hardly be
commensurate with the overall cost. *A fortiori*, this mechanism is oper-
ative to allay such doubts as may be developed by analysts themselves,
who have not only undergone even more arduous training analyses but
have the vested interest of a life-time career commitment and liveli-
hood in the practice of psychoanalysis. And even if analysts waver
nonetheless - as a minority of them do from time to time - there are
other influences at work to quell or mitigate their misgivings.[47] Not
least of these is the intoxication from their socially sanctioned prestige as presumed medical healers of the mind, which is also reflected by
the dependence evinced toward them by their patients.

(d) Freud, who began to publish his ideas towards the end of the
Victorian age, took the obscenity out of sex and helped many to jetti-
son some of the oppressive guilt they had felt about their sexual
urges.[48] Moreover, Freud was an incredibly brilliant, dramatic, capti-
vatingly imaginative and disarmingly persuasive expositor who rested
his case not only on his clinical findings but also on keen perceptions
of daily life which sometimes ring particularly true. Moreover, his
wide learning enabled him to draw on anthropology, literature, and religion, and even on his analyses of the lives of great cultural figures.
As Sutherland[49] put it concisely: "He tries in fact to make sense in
his own terms of the whole of human existence, and there is a temptation to be awestruck by the size of the edifice at the expense of failing to notice whether it is built of bricks or cardboard."

But for the reasons I have given, such motivational considerations
surely do not determine the actual scientific merits of Freudian psy-
chogenics or therapeutics. Indeed, it is conceivable that some day
there just *might* be objectively strong supportive evidence for

Freudianism. Yet even then, most of its fervent adherents may actually
be believing in it for non-evidential emotional reasons. And in that
event, it clearly would yield a correct and enlightening psychological
explanation of their assent to adduce these emotional factors. But,
although this psychological explanation of belief would be illuminating, in the face of the posited evidential support this explanation
could not also justify a psychologistic dismissal of Freudism.[50]

Acknowledgments

The author is greatly indebted to the Fritz Thyssen Stiftung for
the support of research.

This paper is reprinted from *Transactions of the New York Academy
of Sciences*, Series II, vol. 39, 1979, *A Festschrift for Robert Merton*.
It is a much *enlarged* and revised version of the paper by the same
title published originally in *Humanities in Society*, vol. 1, no. 4
(1978), pp. 293-304. Grateful acknowledgment is made to the Center
for the Humanities at the University of Southern California for per-
mission to use this material here.

FOOTNOTES

1 S. Freud, 1925. The resistances to psychoanalysis. In *Collected Papers*. J. Strachey, Ed., Vol. 5 (1959), 173. Basic Books. New York, N. Y.

2 *Ibid.*

3 S. Freud, 1949. *Introductory Lectures on Psychoanalysis* (first published in English in 1922). J. Riviere, Tr., 386-87. Allen & Unwin. London, England.

4 R. Fliess, 1948. Foreword. In *The Psycho-analytic Reader*. R. Fliess, Ed. xv-xviii. International Universities Press. New York, N. Y. I am indebted to Morris Eagle for this reference.

5 *Ibid.*

6 E. Jones, 1946. A valedictory address. *International Journal of Psychoanalysis*, Vol. 27, 11.

7 E. Glover, 1952. Research methods in psychoanalysis. *International Journal of Psychoanalysis*, Vol. 8, 403.

8 *Ibid.*, 408.

9 *Ibid.*, 405.

10 E. Jones, 1955. *The Life and Work of Sigmund Freud*. Vol. 2, 129-134. Basic Books. New York, N. Y. I am indebted to Dr. Edward J. Shoben, Jr. for this reference as well as for the reference in item 13.

11 *Ibid.*, 133.

12 *Ibid.*, 133.

13 H. Nunberg and E. Federn, eds., 1974. *Minutes of the Vienna Psychoanalytic Society*. Vol. III Scientific Meetings #125 (Jan. 4, 1911), #129 (Feb. 1, 1911), and #146 (Oct. 11, 1911). International Universities Press. New York, N. Y.

14 *Ibid.*, 147 (Meeting #129, Feb. 1, 1911).

15 K. M. Colby, 1951. On the disagreement between Freud and Adler. *The American Imago*, Vol. 8, 237.

16 E. Jones, 1955. *Op. cit.* pp. 133-34.

17 *Ibid.*, 134.

18 J. Terner and W. L. Pew, 1978. *The Courage To Be Imperfect, The Life and Work of Rudolf Dreikurs*, pp. 38-39 and 118-119. Hawthorn Books, Inc. New York, N. Y. I am indebted to Edward J. Shoben, Jr. for calling my attention to this work.

19 *Ibid.*, 39.

20 *Ibid.*, 39. This quotation is *part* of a citation for which they
    give the *Journal of Individual Psychology* 20 (1964), 124 as their
    source (on page 46, fn. 30).

21 *Ibid.*, 39.

22 *Ibid.*, 118.

23 *Ibid.*, 118-119.

24 *Ibid.*, 119.

25 *Ibid.*, 125.

26 *Ibid.*, 118.

27 *Ibid.*, 124.

28 A. Grünbaum, 1972. Free will and laws of human behavior. In *New
    Readings in Philosophical Analysis.* H. Feigl, K. Lehrer and W.
    Sellars, Eds., p. 618. Appleton-Century-Crofts. New York, N. Y.

29 A. Grünbaum, 1972. *Op. cit.* pp. 617-618.

30 C. S. Lewis, 1970. *God in the Dock.* Walter Hooper, Ed., p. 275.
    William B. Eerdmans Publishing Co., Grand Rapids, Michigan. I am
    indebted to Rosamond Sprague for this reference.

31 *Ibid.*, 274-277.

32 *Ibid.*, 272.

33 *Ibid.*, 272-273.

34 S. Fisher and R. P. Greenberg, 1977. *The Scientific Credibility
    of Freud's Theories and Therapy.* P. viii. Basic Books, New York,
    N. Y.

35 J. Anderson, 1977. What is Jimmy Carter really like? *Parade, The
    Pittsburgh Press* (Nov. 13), pp. 9-11, Pittsburgh, Pa.

36 S. Freud, 1927. *The future of an illusion.* In *Standard Edition of
    the Complete Psychological Works of Sigmund Freud.* J. Strachey,
    Ed. and Tr. Vol. 21 (1961), pp. 5-56. Hogarth Press, London,
    England.

37 E. Fromm, 1950. *Psychoanalysis and Religion.* P. 12n. Yale Uni-
    versity Press, New Haven, Conn.

38 T. Kiernan, 1974. *Shrinks, etc.* p. 23. Dial Press, New York,
    N. Y.

39 T. Szasz, 1978. *The Myth of Psychotherapy.* (Chapters 7-9) Anchor
    Press, Garden City, N. Y.

40 F. Cioffi, 1970. Freud and the idea of a pseudo-science. In *Explanation in the Behavioural Sciences*. F. Cioffi and R. Borger, Eds., pp. 471-499 and 508-515. Cambridge University Press, Cambridge, England.

41 *Ibid.*, 515.

42 *Ibid.*, 484-485.

43 *Ibid.*, 485.

44 *Ibid.*, 480.

45 A. Salter, 1952. *The Case Against Psychoanalysis*, pp. 11-12. Henry Holt, New York, N. Y.

46 S. Sutherland, 1976. *Breakdown*, p. 111. Weidenfeld & Nicolson, London, England.

47 *Ibid.*, 109-113.

48 A. Salter, 1952. *Op. cit.*, pp. 12-13.

49 S. Sutherland, 1976. *Op. cit.*, p. 116.

50 For a discussion of the present-day scientific merits of Freudian psychoanalysis, see the following:
A. Grünbaum, 1977. How scientific is psychoanalysis? In *Science and Psychotherapy*. R. Stern, L. Horowitz and J. Lynes, Eds., pp. 219-254, Haven Press, New York, N. Y., and 1979. Is Freudian Psychoanalytic Theory Pseudo-Scientific by Karl Popper's Criterion of Demarcation? *American Philosophical Quarterly*, Vol. 16, 131. Also, A. Grünbaum, 1979. Epistemological Liabilities of the Clinical Appraisal of Psychoanalytic Theory. *Psychoanalysis and Contemporary Thought*, Vol. 2. Also, A. Grünbaum, 1980. Epistemological Liabilities of the Clinical Appraisal of Psychoanalytic Theory. *Noûs*, Vol. 14.

PARAPSYCHOLOGY: SCIENCE OR PSEUDO-SCIENCE?

(1) One thing has to be said with emphasis at the start. It is that the case of parapsychology is quite different from most of the others falling within the scope of the Committee for the Scientific Investigation of the Claims of the Paranormal.[1] It is quite different, that is to say, from the factitious, but richly profitable mysteries of the Bermuda Triangle and of the Chariots of the Gods, from astrological prediction, from the extraterrestrial identification of Unidentified Flying Objects, or from most of the other affairs dealt with so faithfully in that committee's useful and entertaining journal *The Sceptical Enquirer*.[2] The crucial difference from these other cases mentioned is that there we either know from the beginning that it is all bunkum, or else we can come to know this very soon after serious and honest investigation has begun.

Thus the moment someone concerned to discover what's what, rather than to produce a best selling real life mystery, began to probe the Bermuda Triangle story it became apparent that there is no sufficient reason to believe that more ships and aircraft vanish without trace in that area than anywhere else with comparable traffic densities and comparable natural hazards. Again, there just is no good reason to believe that there have been any close encounters of the third kind; nor indeed of the first nor second either. The truth here is that the content of visions, dreams, and misperceptions is always in part a function of the wishes, beliefs, and expectations of the subject. So Chinese, under the old Emperors, used to dream dreams of dragons and Confucian officials; but not of Red Guards, chanting doubleplus good Chairman Mao-think. So too Bernadette Soubirois in her nineteenth century French village had a vision of the Blessed Virgin, as represented in pictures and images in her local church; but not of Shiva the Destroyer, as represented in Indian temple sculptures. So, again and likewise, when contemporary North American readers of science fiction misperceive celestial phenomena, what they believe that they have seen is neither gods nor a dragon but a spaceship. Such false identifications are, in one of the finest phrases of Karl Marx, "the illusion of the epoch."

Parapsychology, however, is a horse of quite another color. One
of the properly uncelebrated silver jubilees of 1978 was that of the
publication of my own first book, entitled in English English, with all
the brash arrogance of youth, *A New Approach to Psychical Research*.[3]
Yet it is just worth saying here that, after reviewing the literature
as it then was, I concluded there that, although there was no repeatable experiment to demonstrate the reality of any of the putative psi-
phenomena, and although the entire field was buried under ever-mounting
piles of rubbish produced by charlatans and suckers; nevertheless one
could not with a good academic conscience dismiss the case as closed.
Too much seemingly sound work pointing to the genuineness of at least
some of these phenomena had been done. Too many honest, toughminded,
methodologically sophisticated and often formidably distinguished persons had been involved in this work. Not even the youngest and most
wholehearted of Humians could recommend that we commit it all to the
flames as "containing nothing but sophistry and illusion."[4] The research had to go on.

With, it must be confessed, precious little participation by the
author of "that juvenile work,"[5] the research has indeed gone on. In
all probability its sum in the years between is as great or greater
than the total for all the years before. Yet it is hard to point to
any respect in which the general situation is better now than it was
then. Certainly there is still no repeatable experiment to demonstrate
the reality of any putative psi-phenomenon. Now as then the experts
are inclined to construe the night on night regularity of the performance of any stage or screen psychic as proof that that performance is
nothing but conjuring. Even worse or - according to taste - even better, S. G. Soal's work on Gloria Stewart and Basil Shackleton has been
progressively discredited. This won Soal a D.Sc. from the University
of London, and was hailed by so tough a nut as C. D. Broad as involving,
among other things, "The Experimental Establishment of Telepathic Precognition."[6] Nevertheless, not to put too fine a point on it, Soal,
who was in his later years to present the crudely fraudulent Jones
brothers as *The Mind Readers*,[7] seems to have been faking the scores.[8]

Having so far, in the present Section 1, labored in the main to
distance parapsychology from some wholly disreputable exercises in de-
ception and self-deception, I intend in the remaining three sections to
consider three respects in which it appears to differ from all the
established high-status sciences. First, its field has to be defined

negatively. Second, there is no repeatable demonstration that it does in truth have its own peculiar and genuine data to investigate. And, third, there is no even half-way plausible theory with which to account for the materials it is supposed to have to explain.

(2) In his Gifford lectures, Sir Charles Sherrington remarked that the names given to the vitamins were at first "non-committal in order that scientific ignorance should not be cloaked. Under full knowledge they are already being christened properly and chemically. Vitamin C is ascorbic acid..."[9] It is now usual for parapsychologists to begin by following this excellent example; although here, regrettably, there is no sign of progress towards legitimate rechristenings. 'Parapsychology' is thus defined as 'the study of the psi-phenomena'; 'psi' being the name of the initial letter of the Greek word from which our 'psychic' is derived. Psi-phenomena are divided into two fundamental categories: psi-gamma; and psi-kappa. 'Gamma' and 'Kappa' are again names for Greek letters; the initial letters, respectively for the Greek words for knowledge and movement. The word 'psi-gamma' covers both of what are elsewhere more tendentiously described as clairvoyance (clear seeing) and telepathy (distant feeling). The word 'psi-kappa' substitutes for the equally tendentious 'psychokinesis' (movement by the mind).

(i) We speak, or would speak, of psi-gamma when some subject comes up with information; and when that subject's acquisition of this information cannot be put down either to chance, or to perception, or to inference from materials ultimately obtained through sensory channels. These phenomena, or alleged phenomena, are then subdivided in two ways. One distinction is between clairvoyant and telepathic conditions. The idea is to distinguish two kinds of psi-gamma information: that already available to some person other than the subject, and presumably being somehow acquired from that other person; and that not available to any other person but immanent in the non-personal world, and presumably being somehow acquired directly from that non-personal world. The tradition, strongly challenged yet dominant still, takes a Platonic-Cartesian view of the nature of man for granted. So it describes the former as mind to mind, the latter as matter to mind.

The other distinction refers to temporal order. If the information produced in or by the subject is only going to become normally available in the future, then it is usual to speak of paranormal

precognition or of precognitive psi-gamma. With appropriate alterations the same formula will give the meanings of 'paranormal retrocog-
nition' and 'retrocognitive psi-gamma.' When there is no such qualify-
ing adjective we may take it that the psi-gamma is neither precognitive
nor retrocognitive but simultaneous.

Once these several definitions are given and understood, it must
become immediately obvious that it is inept - not to say perverse - to
characterize psi-gamma as a new form of either perception or knowledge.

(a) If the word 'extra' in the expression 'Extra-
Sensory Perception (ESP)' is construed as meaning outside of - like the
'extra' of 'extra-marital sex' - then that expression becomes self-
contradictory. It becomes equivalent to 'extra-perceptual perception';
and hence, as Thomas Hobbes would have had us add, parallel to 'incorporeal substance.' If, on the other hand, 'extra-sensory' is interpreted as referring to an hypothetical additional sense, then that hypothesis is at once falsified by two decisive deficiencies. First,
there is no bodily organ or area the masking or local anaesthetization
of which suppresses psi-gamma. Second, there is no accompanying sixth
mode of sensory experience as different from visual, tactual, gusta-
tory, auditory, and olfactory as each of these is different from all
the others. For good measure we may conclude the paragraph by mentioning a further deficiency. It seems that the subjects who come up with
the information are unable at the time to recognize the deliverances of
this supposed new sense, and to distinguish them from plain ordinary
guesses or hunches or imaginings.

(b) Since psi-gamma information is defined as precisely
not being acquired through the senses it really is, as has just been
urged, perverse to insist upon thinking of psi-gamma in terms of a
perceptual model. It is almost equally perverse to think of such information as constituting a kind of knowledge. For the definition
stipulates that the subject must not be in a position to know, either
on the basis of perception, or on the basis of inference from antece-
dently available material. If, but only if, subjects were at the time
of coming up with the information able to pick out some of the items
as coming from a fresh, special and reliably veridical source; then indeed we might quite properly begin to speak of belief in the truth of
these items as knowledge, knowledge duly grounded in that source or
faculty. But that ability is no part of the accepted definition of
'psi-gamma.' Nor would it be sensible to require it by adding a

further clause. For it appears that such ability is rather seldom
claimed, and never in fact found, among those responsible for what are,
on the established weaker definition, ostensible cases of psi-gamma.[10]

Since the suggestion that we have here a fresh form of knowledge
is, for the reasons given, wrong, I regret that no one took up my pro-
posal to make the temporal distinctions by applying to the Greek noun
'psi-gamma' the appropriate member of a trio of more familiar Latin
letters: M (for minus, replacing retrocognitive); S (for simultaneous);
or P (for plus, replacing precognitive).[11]

(ii) For present purposes the most important feature of the
definitions presented in Subsection (i) is that they stipulate what
psi-gamma is not, rather than what it is, or would be. In the opening
words of one especially thoughtful Presidential Address to the Society
for Psychical Research (London): "The field...must be unique in one
respect at least: no other discipline, so far as I know, has its subject matter demarcated by exclusively negative criteria. A phenomenon
is, by definition, paranormal if and only if it contravenes some fundamental and well-founded assumption of science."[12]

(a) There is, I imagine, no disputing but that this
must make it harder to establish that there really is psi-gamma: the
difficulty of proving negatives is notorious and trite. But some other
consequences are less obvious and more disputatious. Take first the
points of Subsubsection 1 (i) (b), above, and especially the last two:
that subjects are not able to pick out items as coming from a fresh,
special, and reliably veridical source; and that it neither is nor
ought to be part of the meaning of 'psi-gamma,' that they should be so
able.

From all this it surely follows, as is indeed the case, that psi-
gamma can only be identified by subsequent checkups; and hence that it
is not an independent source of knowledge. Thus, in the experimental
work, the only way of telling whether or not we have any psi-gamma effect is by scoring up the subjects' guesses against the targets, and
then calculating whether the proportion of hits to misses is too great
to be dismissed as no better than what could have been expected 'by
the law of averages.'

The case is substantially the same with what Broad would have us
describe as, rather than spontaneous, sporadic psi-phenomena. There
is again no way of identifying information coming telepathically or
clairvoyantly save by comparing the hunches, dreams, visions, thoughts

or what have you of the subject with whatever it is to which they may
or may not correspond; and then estimating as best we can whether or
not the degree of correspondence is greater than might reasonably be
put down to chance, perception, or conscious or unconscious inference
from materials ordinarily available to the subject.

The conclusions that psi-gamma as at present defined can only be
identified by subsequent sensory checkups, and that it is therefore not
an independent source of knowledge, carry an interesting corollary.
This corollary seems to have been noticed only once or twice, and never
discussed.[13] It is that, even supposing that we were able to construct
a coherent concept of an incorporeal soul surviving the dissolution of
its body, we could not consistently suggest that such souls might first
learn of one another's existence, and then proceed to communicate,
through psi-gamma.

For suppose first that there were such incorporeal Cartesian subjects of experience. And suppose further that there is from time to
time a close correspondence between the mental contents of two of
these beings; although such a fact could not, surely, be known by any
normal means to anyone in either our world or the next. Now, how could
either of these two souls have, indeed how could there be, any good reason for hypothesizing the existence of the other; or of any others? How
could such beings have, indeed how could there be, any good reason for
picking out some of their own mental contents as - so to speak - mes-
sages received; for taking these but not those to be, not expressions of
a spontaneous and undirected exercise of the imagination, but externally
provoked communication input? Suppose these two challenging questions
could be answered, still the third would be "the killing blow." For
how could such beings identify any particular items as true or false,
or even give sense to this distinction?

The upshot seems to be that the concept of psi-gamma is essentially
parasitical upon everyday, this-worldly notions; that, where there could
not be perception, there could not be 'extra-sensory perception' either.
It is assumed too often and too easily that psi-capacities not only can
be, but have to be, the attributes of something immaterial and incorporeal; mainly for no better reason than that they would be non-physi-
cal in the quite different sense of being outwith the scope of today's
physical theories. Yet the truth is that the very concepts of psi are
just as much involved with the human body as are those of other human
capacities and activities. In the gnomic words of Wittgenstein: "The
human body is the best picture of the human soul."[14]

(b) A second important, too rarely remarked consequence of the fact that the definition of 'psi-gamma' is negative, is that the concept itself, and not just the best available evidence that it does in fact have some application, is essentially statistical. Consider first a standard experiment in which a subject guesses through a well-shuffled pack of Zener cards - five suits of five identical cards - while an agent, suitably concealed from the subject, exposes to himself, and briefly contemplates, each card in turn. We are, that is, supposing telepathic as opposed to clairvoyance conditions. And suppose that, after this procedure has been many times repeated, it emerges that the subject has scored significantly better, or worse, than the expected chance average rate of one in five. Then on the face of it we have a case of psi-gamma.

But now notice that we have absolutely no way of picking out from the series any single hit, or any collection of particular hits, and identifying this, or these, as due to the subject's psi-capacity, rather than to chance. Or, rather, that is misleading. It is not that we as a matter of fact at this time cannot thus divide the singly paranormal from the singly normal. The crux is that no meaning has been given to this distinction: psi-gamma just is understood as a factor which manifests itself, if at all, only in the occurrence of significant deviations from mean chance expectation over a series of guesses - or over a series of whatever else it may be.

This is also one of the reasons why it is misleading to speak of a subject who puts up a score significantly better, or worse, than mean chance expectation as doing this *by* or *by means of* telepathy, clairvoyance, or other paranormal power. For while it remains possible, at least as far as the present consideration goes, for theorists to hypothesize some so far unrecognized kind of radiation through which information is conveyed to subjects; still 'psi-gamma' is at this time defined as precisely not the product of any means we can think of.[15] If the subject used the methods of the conjuror, or cheated by stealing a peek at the target cards, or had some hand in the determination of their values; then the results are on these grounds disqualified as not genuine psi-gamma.

Some have thought to dismiss the contention of this subsubsection on the ground that it does not apply to sporadic psi-gamma. But it does. Consider, for instance, the person who - "on the night when that great ship went down" - had a dream which both they and the

parapsychologists are inclined to rate as telepathic. Their case will
rest, not upon any particular correspondence between the dream images
and the reality, but upon the total amount of that correspondence. A
perfect fit of the whole would be the sum of fits at every particular
point. Once again, there is no way of determining, and no sense in
asking, which of these particular fits should be scored to chance and
which to psi-gamma.

(iii) In this paper I am concentrating on psi-gamma, without
asking systematically how much of what I say applies to psi-kappa. But
I cannot leave the point about the essentially statistical character of
the former without - not for the first time - drawing attention to a
most remarkable fact. The fact is that all the evidence for the latter, and almost all the work on it, is similarly statistical. Yet the
concept of psi-kappa is not. In the Glossary printed in every issue of
the *Journal of Parapsychology* 'psi-kappa (PK)' is defined as "the direct influence exerted on a physical system by a subject without any
known intermediate physical energy or instrumentation." More popularly, it is the putative power to move something, or at least to impress
a force upon it, by just willing; and without touching it or employing
any electrical or mechanical device to bring this result about.

Now, it should be immediately obvious that there is no analogue
here for those mere chance correspondences which investigators of psi-
gamma through their statistical calculations labour to discount. There
seems to be no a priori reason why psi-kappa should have to be detected
and studied as the production, by a subject just 'willing,' of a significant surplus of sixes among the falls of dice mechanically rolled ten
or more at a time; rather than as the production, by the same subject
just 'willing,' of particular single movements in some highly sensitive
and scrupulously shielded physical instrument. On the contrary: it
would seem a priori far more likely that subjects would be able to direct their - shall we say? - willpower at a single stationary target
than at (presumably at most one or two of) the several dice moving
rapidly yet raggedly in midair. For would not such direction require a
find-fix-and-strike mechanism comparable with what is needed in an an-
tiballistic missile (ABM) defence system?

Of course, nature neither has to be nor is slave to our notions of
the a priori probable or improbable. So it may be that in fact it is
easier or only possible to deploy the force - May the force be with
you, investigators! - against either a confusion of dice spinning in

mid air or a jostling mob of paramecia. But the observation of the
present subsection must still raise questions about experimenters who
seem never in the first decade or so, and rarely later, either to have
effected tests of the most obvious kind, or to have provided any rationale for their long-sustained refusal so to do.[16]   Rhine, I believe,
spoke truer than he either knew or would have cared to know when, in
1947, he insisted: "The most revealing fact about PK is its close tie
up with ESP...".[17] For the uncanny resemblances between, on the one
hand, the methods and findings of the experimental investigation of
psi-gamma and, on the other hand, those of psi-kappa, do in truth constitute strong, though much less than decisive, reason for concluding
that what we have in both cases is evidence: not so much of some previously unrecognized personal power; but rather of a lot of fraud,
self-deception or incompetence - and maybe of some real statistical
oddities not significant of causal connections. There certainly is
'Something Very Unsatisfactory' about what is supposed to be evidence
of putative personal power, yet in which there seem to be no close concomitant variations between the effects alleged and any psychological
variables in the supposed effectors.[18]

     (3) One of the most important similarities between the two main
subareas of the field of parapsychological experimentation is that,
typically, the work of one investigator cannot be repeated by another:
not even when the second is able to use the same subjects as the
first.[19] This fact is one of several which give purchase for the representation here of Hume's once notorious arguments about the difficulty, amounting usually to the impossibility, of establishing upon
historical evidence that miracles have occurred.[20] These arguments
were thus a few years ago redeployed in the present context by G. R.
Price.[21] This Price, it has to be said, must not be confused with two
others better known in this field: the late disreputable Harry Price,
who surely faked some of the Borley Rectory phenomena which he was pretending to investigate; and the most excellent sometime Oxford professor, well known for an almost Kantian integrity.

     Hume, it will be remembered, contended "that no testimony for any
kind of miracle has ever amounted to a probability, much less to a
proof; and that even supposing it amounted to a proof, it would be
opposed by another proof; derived from the very nature of the fact,
which it would endeavor to establish." Confronted by such a conflict

of evidence, and - the interpreter must interject - remembering Hume's
unfortunate ambition to develop a psychological mechanics, "we have
nothing to do but subtract the one from the other, and embrace an opinion, either on one side or the other, with that assurance which arises
from the remainder." However, for reasons which are not made altogether clear, "this subtraction, with regard to all popular religions,
amounts to an entire annihilation; and therefore we may establish it as
a maxim, that no human testimony can have such force as to prove a
miracle, and make it a just foundation for any such system of religion."[22]

　　　　(i) A miracle for Hume would be much more than a fact
"which...partakes of the extraordinary and the marvellous." For, by
the force of the term, "A miracle is a violation of the laws of nature..." (A footnote adds a supplementary clause: "A miracle may be
accurately defined, *a transgression of a law of nature by a particular
volition of the Deity, or by the interposition of some invisible
agent*.")[23] Waiving on this occasion the scholarly question whether Hume
himself was in any position to provide an account of laws of nature
strong enough to permit this contrast between a miracle and a fact
which merely "partakes of the extraordinary and the marvellous," we
need first to show that reports of psi-phenomena would lie within the
range of Hume's argument.

　　　　　　(a) There is no doubt but that they would, or do. Certainly it is not easy to think of any particular named law of nature -
such as Boyle's Law or Snell's Law or what have you - which would be,
or is, as Hume would have it, 'violated' by the occurrence of psi-gamma
or psi-kappa. What that threatens is more fundamental. For the psi-
phenomena are in effect defined in terms of the violation of certain
"basic limiting principles"; principles which constitute a framework
for all our thinking about and investigation of human affairs, and
principles which are continually being verified by our discoveries.
If, for instance, official secret information gets out from a government office, then the security people try to think of every possible
channel of leakage: and what never appears on the check lists of such
practical persons is psi-gamma. When similarly there has been an explosion in a power station or other industrial plant, then the investigators move in. At no stage will they entertain any suggestion that no
one and nothing touched anything, that the explosion was triggered by
some conscious or unconscious exercise of psi-kappa. Nor shall we

expect them to turn up any reason for thinking that their, and our, framework assumptions were here mistaken.

It is some of these usually unformulated "basic limiting princi-ples" which both psi-gamma and psi-kappa would, or do, violate; and which C. D. Broad formulated in his much reprinted *Philosophy* article on "The Relevance of Psychical Research to Philosophy."[24] Broad's for-mulations here are pervasively Cartesian. They thus provide for "the interposition of some invisible agent, if not for a particular voli-tion of the Deity." What, for instance, psi-kappa would violate is the principle that "It is impossible for an event in a person's mind to produce directly any change in the material world except certain changes in his own brain...it is these brain-changes which are the immediate consequences of his volitions: and the willed movements of his fingers follow, if they do so, only as rather remote causal descendants."[25]

A Rylean, of course, would attribute any psi feats to the flesh and blood person rather than to his putative incorporeal mind or soul. But Broad, taking absolutely for granted a fundamentally Cartesian view of the nature of man, is instead so misguided as to conclude that it is the supposed establishment of the reality of the psi-phenomena, rather than this unnoticed and unargued preconception, which "has undermined that epiphenomenalist view of the human mind and all its activities, which all other known facts seem so strongly to support..."[26]

(b) In their first response to G. R. Price's paper "Science and the Supernatural" Paul Meehl and Michael Scriven wrote: "Price is in exactly the position of a man who might have insisted that Michelson and Morley were liars because the evidence for the physical theory of that time was stronger than that for the veracity of these experimenters."[27] It is important to appreciate why this is not so. Two of the reasons I shall consider here and now; the third is the sub-ject of Section 4, below.

First, the Michelson-Morley experiment was not one in a long ser-ies including many impressively disillusioning instances of fraud and self-deception. Second, there was in that case no reason at the time - nor has any reason emerged since - for suspecting that the experiment would not be repeatable, and repeated; as well as confirmed indirectly by other experiments similarly repeatable, and repeated. It is these two weaknesses together which lay parapsychology wide open to the Humi-an challenge, each weakness reinforcing the other. The black record of fraud would not carry nearly so much weight against what might seem to

be strong new cases of psi; if only we possessed some repeatable demonstration of the reality of such phenomena. We should not be in such
desperate need of that repeatable demonstration; if only there had not
been so much fraud and self-deception.

This is perhaps the moment to perform the nowadays mandatory genu-
flexion towards Thomas Kuhn's *The Structure of Scientific Revolutions*.
Normal science - here to be construed as contrasting with pseudo-
science rather than science in revolution - involves "research firmly
based upon one or more past scientific achievements...that some particular community acknowledges for a time as supplying the foundation for
its further practice."[28] Such an acknowledged achievement, if the ac-
knowledgement and the diploma title 'science' are to be deserved, must
surely embrace some measure of demonstrable repeatability. For - becoming now a brazen and reactionary non-Kuhnian - remember that the aim
of science is, after discovering what sorts of things happen, to explain why: the qualification 'sorts of' has to go in to cover the
point that, unlike history, science is concerned with the type rather
than the token. The formula for the repetitive production of a type is
at the same time an initial, no doubt inadequate, explanation of the
occurrence of any and every particular token of that general type;
while in these two aspects together the achievement of that formula
constitutes a pledge of more and better yet to come.

So, until and unless the parapsychologists are able to set up a
repeatable demonstration, they will at best be making preparations for
the future development of a future science - with no guarantees that
these aspirations ever will in fact be realized. One moral to draw
from this point, and indeed from the whole paper in which it is made,
is that, if affiliation to the AAAS is thought of as a recognition of
actual achievement rather than of good intentions, then the Parapsychological Association is not yet qualified for admission, and ought
now to be politely disaffiliated.

(c) It is sometimes suggested, either that repeatability does not matter; or else that there already is as much of it in
parapsychology as there was in say the study of magnetism before elec-
tricians learnt how to construct artificial magnets, or as there is now
in abnormal psychology.[29] But these analogies break down. Certainly
alleged star performers in psi-gamma or psi-kappa are, like natural
lodestones or calculating boys, rare. But, when the latter are found,
different investigators regularly repeat the same results. The same,
unfortunately, is not true with psi.

Nor will it do to dismiss the demand for repeatability as arbitrary or unreasonable. For, if only it could be satisfied, then parapsychology would escape the Humian challenge. But, as it is, any piece of work claiming to show that psi-phenomena have occurred is in effect a miracle story. So, in order to form the best estimate we can of what actually happened, we have to resort to the methods of critical history. This means that we have to interpret and assess the available evidence in the light of all we know, or think we know, about what is probable or improbable, possible or impossible. But now, as we saw earlier in the present subsection, psi-phenomena are implicitly defined in terms of the violation of some of our most fundamental and best evidenced notions of contingent impossibility. So, even before any Humian allowance is made for the special corruptions afflicting this particular field, it would seem that our historical verdict will have to be, at best, an appropriately Scottish, and damping: 'Not proven.'

      (ii) Hume started with a general argument about the difficulty of establishing upon historical evidence the occurrence of a miracle. He then proceeded to contend that this difficulty is compounded when the miracle stories in question have "regard to...popular religions." So much so that he felt entitled to conclude "that no human testimony can have such force as to prove a miracle, and make it a just foundation for any such system of religion."

Whatever force Hume's wordly contentions here may have must bear equally against the miracle stories of parapsychology. For the Founding Fathers unanimously believed that to establish the reality of what we now call the psi-phenomena would be to refute philosophical materialism; thus opening the way to an empirically grounded doctrine of personal survival, even personal immortality. Frederic Myers, for instance, in his 1900 Presidential Address to the original Society for Psychical Research (London) said, in as many words, that their goal was to provide "the preamble of all religions," and to become able to proclaim: "thus we demonstrate that a spiritual world exists, a world of independent and abiding realities, not a mere 'epiphenomenon' or transitory effect of the material world."[30]

Again, Henry Sidgwick in his own second Presidential Address, speaking of the motives of the whole founding group, explained how "it appeared to us that there was an important body of evidence - tending *prima facie* to establish the independence of soul or spirit...evidence tending to throw light on the question of the action of mind either

apart from the body or otherwise than through known bodily organs."[31]
In his third Presidential Address he added: "There is not one of us
who would not feel ten times more interest in proving the action of
intelligence other than those of living men, than in proving communication of human minds in an abnormal way."[32]

In our own day J. B. Rhine's best-selling accounts of the research
at Duke University present it all as proving some sort of Cartesian
view of the nature of man, and refuting philosophical materialism.
"The thread of continuity," he writes, "is the bold attempt to trace as
much as we can see of the outer bounds of the human mind in the universe."[33] Descriptions of familiar flesh and blood creatures guessing
cards, or 'willing' dice to fall their way, are spiced with references
to minds; their powers, frontiers, and manifestations; their unknown,
delicate and subtle capacities; and the experimental findings are all
construed as striking hammer blows for 'spiritual values' in the global
battle against 'materialism.' Always Rhine deplores "the traditional
disinclination to bring science to the aid of our value system."[34]

It is, by the way, a noteworthy indication of the enormous power
and fascination of the Cartesian picture that, as we have seen, even so
acute and so unspiritually-minded a philosopher as Broad took it as obvious that to establish the reality of paranormal human powers is, both
to establish the reality of incorporeal thinking substances as the
bearers of those powers, and to undermine the plausibility of the epiphenomenalist account of the relation between consciousness and the conscious organism. Yet what reason did Broad have for attributing these
putative powers to such unidentified and unidentifiable metaphysical
entities,[35] rather than to those familiar flesh and blood creatures who
to the philosophically uncontaminated eye are the ostensible performers?

(4) In Section 1, I distinguished, as the third peculiarity pre-
judicing the scientific pretensions of parapsychology, the fact that
"there is no even half-way plausible theory with which to account for
the materials it is supposed to have to explain." This deficiency
bears on the question of scientific status in two ways. For a theory
which related the putative psi-phenomena to something else less conten-
tious would tend: both to probabilify their actual occurrence; and to
explain why they do thus indeed occur. Here we have the third reason
why to refuse to accept the reality of such phenomena is not on all
fours with dismissing the result of the Michelson-Morley experiment.

For, even if no one then was ready immediately with an alternative
theory, still in that case there was no good reason to fear that such
a theory could not be produced. But, in the case of parapsychology
now, our investigators have had getting on for a hundred years for
theoretical cogitation, while there is also reason to believe that at
least some of the phenomena alleged are so defined as to be necessarily
impervious to causal explanation.

     (i) The situation is confused by the fact that most investi-
gators have been, and are, attached to a conceptual scheme whose actual
explanatory power they tend vastly to exaggerate. For, as we have
seen, most of them, taking the Cartesian concept of soul to be quite
unproblematic, are ready to construe any proof of the reality of psi-
phenomena as at the same time proof of the existence and activities of
Cartesian souls. So, as they come to believe in this reality, they
forthwith attribute all such performances to those putative agents.
When they leap to this congenial conclusion: not only do they overlook
the by now surely notorious difficulties of offering any serviceable
description to enable these incorporeal somewhats to be identified, in-
dividuated, and reidentified through time; they also fail to provide
their proposed hypothetical entities with any characteristics warrant-
ing the expectation that these could, and naturally would, achieve what
for mere creatures of flesh and blood must be simply impossible. If,
as C. W. K. Mundle put it in his 1972 Presidential Address to the Society for Psychical Research (London), "materialism is to be rejected
in favor of dualism on the ground that materialism cannot explain all
kinds of ESP, it needs to be shown that, and how, all kinds of ESP can
be explained in terms of immaterial minds."[36]

    Where detail is vouchsafed sufficient to yield a piece of discus-
sable and even testable theory, the result is almost if not quite always a fragment; a fragment which could at best serve only to explain
one kind of psi-phenomenon - simultaneous psi-gamma under telepathic
conditions. This applies, for instance, both to Whately Carington's
proposals about the association of ideas[37] and to Ninian Marshall's
physicalistic postulation of an assimilative force by which all physical things tend to make others more like themselves.[38] But the evidence
for straight simultaneous psi-gamma under telepathic conditions now
appears to be neither substantially stronger than, nor of a significantly different kind from, the evidence for 'precognitive' psi-gamma,
simultaneous psi-gamma under clairvoyance conditions, or psi-kappa.

So it looks as if our choice is: either to think up a comprehensive
theory covering all kinds of psi-phenomena, and presumably a lot else
besides; or else to go back in the end to the position which we could
not in good academic conscience adopt at the beginning - that of committing the whole pseudo-subject to the flames, in high Humian style, "as
containing nothing but sophistry and illusion."

    (ii) Already in Subsection 2 (iii) I hinted at the great ob-
stacle in the way of an explanation of psi-kappa: this is the problem
of describing some believable 'find-fix-and-strike' mechanism for directing the force "at (presumably at most one or two of) the several
dice moving rapidly yet raggedly in mid-air"; and when those dice or
other larger objects may not be with the sensory range of the 'willer.'
We have now to notice in passing a similar massive obstacle standing in
the way of any attempt to explain psi-gamma under clairvoyance conditions. Suppose that someone does spectacularly better than mean chance
expectation in guessing the values of the cards in a well-shuffled
pack, guessing these "down through"; with no one touching that pack until the complete guess-run is later scored. What conceivable mechanism
could that subject have employed - unconsciously, of course - to acquire the information needed to achieve such scores? (An appreciation
of the force of this question has held many psychical researchers back
from accepting clairvoyance even when they have no remaining doubts
about telepathy.)

    (iii) These are both formidable difficulties for the specula-
tor. But the obstacle barring the way to any explanation which accepts the genuineness of P psi-gamma (precognition) is of an altogether
different kind, and totally decisive. For if the various conditions
usually specified as essential to a genuine case of P psi-gamma are in
truth all satisfied simultaneously, then what is going on just is not
susceptible to explanation in terms of causes or of causally interpret-
able natural laws. It is significant that when Broad offered a theory
'to explain precognition' he added this warning: "...notice that, on
this theory of 'precognition,' no event is ever 'precognized' in the
strict and literal sense."[39]

The crux is that inexplicability is built into what is here rather
oddly called "the strict and literal sense" of (paranormal) precognition. For consider, to start with we must have highly significant correlations between what someone says, or does, or experiences and what
is later said, or done, or happens. But though necessary this is by

no means sufficient. If the subject played any part in bringing about
those later ongoings, then that would be enough to disqualify that subject's anticipations as a case of P psi-gamma. Suppose next that the
correlation between these anticipations and their later fulfillments
could be explained in terms of some common causal ancestry: maybe the
guesser is an identical twin of the person choosing the targets, and
they share genetically determined patterns of guessing and choosing dispositions. Here too the correlations would be disqualified: it is indeed mainly in order to avoid disqualification by reference to a common
causal ancestry that experimenters insist that the targets must be randomly selected.

But now, what possibility of causal explanation is left? If anticipations and fulfillments are causally connected, then either the anticipations must cause the fulfillments, or the anticipations and the
fulfillments must both be partly or wholly caused by something else, or
the fulfillments must cause the anticipations. The first two disjuncts
are, as we have just seen, ruled out by the force of the (expression)
term (paranormal) 'precognition.' The third is radically incoherent.
Because causes necessarily and always bring about their effects, it
must be irredeemably self-contradictory to suggest that the (later) fulfillments might cause the (earlier) anticipations. By the time the fulfillments are occurring the anticipations already have occurred. It
would, therefore, be futile to labour either to bring about or to undo
what is already unalterably past and done.[40]

I indicated in the first paragraph of this final Section 4 how a
well-supported theory may probabilify the occurrence of whatever it pre-
dicts. So I trust that it will not look like a lapse into anti-empiri-
cal dogmatism to conclude with a maxim from Sir Arthur Eddington, a
leading British physicist of the period between the wars: "It is also
a good rule not to put overmuch confidence in the observational results
until they are confirmed by theory."

FOOTNOTES

1   This was set up a year or two ago on the initiative of Professor
    Paul Kurtz of SUNY at Buffalo, the then Editor of *The Humanist*, in
    hopes of doing something to stem the rising tide of popular creduli-
    ty.

2   Formerly The *Zetetic*, now edited by Kendrick Frazier from 3025 Palo
    Alto Drive, NE, Albuquerque, New Mexico, 87111.

3   London: C. A. Watts, 1953. This book was long ago remaindered,
    and its publishers were later absorbed into the fresh-founded firm
    Pemberton Books. But two chapters, revised, are still current in
    philosophical anthologies. See, for one, Note 34.

4   *An Inquiry Concerning Human Understanding*, XII (iii), *ad. fin.*

5   See the Advertisement, referring to the *Treatise*, added by Hume to
    what in the event became the first posthumous edition of that *In-
    quiry*.

6   Compare his article under this title in *Philosophy* (London), Vol.
    XIX (1966), 261-75. Broad was only perhaps the most distin-
    guished of the many who accepted this work at its face value. That
    throng - or should I say that rout? - included the author of *A New
    Approach to Psychical Research*.

7   London: Faber and Faber, 1959.

8   See, most recently, Betty Markwick "The Soal-Goldney Experiments
    with Basil Shackleton: New Evidence of Data Manipulation," in *Pro-
    ceedings of the Society for Psychical Research* (London), Vol. LVI
    (1978), 250-78. Compare D. J. West "Checks on ESP Experimenters,"
    in the *Journal* of the same society, Vol. XLXIX (1978), 897-9; also
    further references given in these articles and in Edward Girden
    "Parapsychology," Chapter 14 of E. C. Cartette and M. P. Fried-
    man (Eds.) *Handbook of Perception*, Vol. X (New York: Academic
    Press, 1970), 385-412, especially pp. 396 ff. All this sheds a
    very bright light upon Soal's "calm, but perfectly devastating
    reply" to B. F. Skinner's contention that mechanical scoring de-
    vices should have been employed, to reduce the possibilities of
    cheating. (See G. R. Price "Where is the Definitive Experiment?"
    reprinted in J. Ludwig (Ed.) *Philosophy and Parapsychology* (Buffa-
    lo, N. Y.: Prometheus, 1978), 197-8).

9   *Man on His Nature* (Cambridge: CUP, 1946), p. 96.

10 See, for instance, C. J. Ducasse in J. Ludwig (Ed.) *loc. cit.*, p.
    131.

11 Almost everyone embarking on a discussion of 'the philosophical
    implications of (paranormal) precognition' has in fact got off on
    the wrong foot by first thinking of P psi-gamma on the model of
    either cognition or perception or both. See, for instance, my

article "Precognition" in Paul Edwards (Ed.) *The Encyclopaedia of Philosophy* (New York: Macmillan and Free Press, 1967), Vol. I, 139-50; or, more full, my "Broad and Supernormal Precognition," P. A. Schilipp (Ed.) *The Philosophy of C. D. Broad* (New York: Tudor, 1959), 411-35.

12 John Beloff, reprinted in J. Ludwig (Ed.) *loc. at.*, p. 356: the first printing was in the Society's Journal, Vol. XLII (1963), 101-

13 See my "Is there a Case for Disembodied Survival?" in the *Journal of the American Society for Psychical Research*, Vol. LXVI (1972), 129-44; perhaps more easily found in J. M. O. Wheatley and H. L. Edge (Eds.) *Philosophical Dimensions of Parapsychology* (Spring-field, Illinois: C. C. Thomas, 1976) or - in a revised and reti-tled version - in my own *The Presumption of Atheism* (New York: Barnes and Noble, 1976).

14 *Philosophical Investigations*, tr. G. E. M. Anscombe (Oxford: Blackwell, 1953), p. 278.

15 So far as I know the first person to make much of this point that psi-gamma is essentially meansless was Richard Robinson. See his contribution to the Symposium "Is Psychical Research Relevant to Philosophy?", originally published in the *Proceedings of the Aris-totelian Society*, Supp. Vol. XXIV (1950), but reprinted in J. Ludwig (Ed.) *loc. cit.* The relevant passage is at pp. 80-3 in that reprinting.

16 Compare my 'Something Very Unsatisfactory,' in the *International Journal of Parapsychology*, Vol. VI (1964), 101-5.

17 See E. Girden "A Review of Psychokinesis (PK)," in the *Psychological Bulletin*, Vol. LIX (1962), at pp. 353 ff.

18 Compare Note 16, above. The only apparent exception to the gener-alization in the text is the sheep/goat work initiated by Dr. Ger-trude Schmeidler: the 'sheep,' who believe in the reality of psi-phenomena, usually put up better scores than the 'goats,' who do not. But, even if the findings here were unequivocal, they would not provide the sort of dramatic concomitant variation reported in - and, apparently, reported only in - the since discredited work of Soal: there when the conditions were suddenly and secretly changed from telepathy to clairvoyance one subject's scores at once dropped to a chance level, recovering equally immediately when the original experimental conditions were restored.

19 Compare, again, Girden *loc. cit.*

20 I develop and defend an interpretation of these arguments in Chap-ter VIII of my *Hume's Philosophy of Belief* (New York: Humanities Press, 1961). So here I follow without defending that same inter-pretation.

21 This Humian challenge was first published in *Science*, Vol. CXXII, No. 3165 (1955), 359-67; but it is now more easily found, along with one or two counterblasts, in Ludwig *loc. cit.*

22 *An Inquiry concerning Human Understanding*, X (ii); p. 127 in the
    standard Selby-Bigge edition. I suppose that "the principle here
    explained" - a principle which so remarkably ensures that 'sub-
    straction' of the lesser quantity from the greater neither leaves
    the greater undiminished, nor diminishes it by the amount 'sub-
    stracted,' but instead always yields a zero remainder - is the
    sum of all the considerations, which Hume has been deploying in
    Part II of this Section X, for thinking that the evidence for the
    occurrence of miracles, which, if they occurred, would support
    "popular religions," is peculiarly rubbishy and corrupt. The up-
    shot should be that it is all this superstitious evidence which is
    thus annihilated, and hence that there is here nothing to 'sub-
    stract' from the strong contrary evidence that whatever laws are in
    question do in fact obtain universally. But this is not what Hume
    actually wrote; which is, it has to be admitted, just muddled.

23 *Ibid.*, pp. 113, 114, and 115*n*: italics original.

24 *Philosophy* for 1949 (Vol. XXIV), 291-309; reprinted in J. Ludwig
    loc. cit., 43-63.

25 *Ibid.*, p. 46: italics original. This principle is Broad's "(2)
    Limitations on the Action of Mind on Matter."

26 *Ibid.*, p. 63; and perhaps compare my *A Rational Animal* (Oxford:
    Clarendon, 1978).

27 J. Ludwig *loc. cit.* 187-8.

28 (Chicago: University of Chicago Press, 1962), p. 10.

29 For the second of these see M. Scriven "The Frontiers of Psycholo-
    gy," in R. G. Colodny (Ed.) *Frontiers of Science and Philosophy*
    (Pittsburgh: University of Pittsburgh Press, 1962), reprinted in
    Wheatley and Edge (Eds.) *loc. cit.*, 46-75. The relevant paragraphs
    are in this reprint at 64-5.

30 *Proceedings of the S. P. R.*, Vol. XV, 117.

31 *Ibid.*, Vol. V, 272-3.

32 *Ibid.*, Vol. V, 401.

33 *The Reach of the Mind* (London: Faber, 1948), p. 50.

34 *Telepathy and the Human Personality* (London: Society for Psychi-
    cal Research, 1950), p. 36. For critiques of such Cartesian mis-
    constructions compare: first, my "Minds and Mystifications," in
    The Listener, Vol. XLVI (1951), 501-2, reprinted in P. A. French
    (Ed.) *Philosophers in Wonderland* (St. Paul, Minn.: Llewellyn,
    1975), 163-7; and, second, "Describing and Explaining," in the
    book mentioned in Note 3, revised and reprinted in J. Ludwig (Ed.)
    loc. cit., pp. 207-27.

35 Compare Terence Penelhum "Survival and Disembodied Existence," a
    potpourri from his book under the same title, cooked up specially
    for inclusion in Wheatley and Edge *loc. cit.*, 308-29.

36   "Strange Facts in Search of a Theory," in Wheatley and Edge (Eds.)
     loc. cit. 76-97:   the sentence quoted is at p. 88.

37   *Matter, Mind and Meaning* (London: Methuen, 1949), 203 ff.

38   "ESP and Memory: A Physical Theory," in the *British Journal for
     the Philosophy of Science*, Vol. X (1960), 265-86.

39   *Religion, Philosophy and Psychical Research* (London: Routledge and
     Kegan Paul, 1953), 80.

40   Compare the unusually hard-hitting symposium "Can an Effect Precede
     its Cause?", in *Proceedings of the Aristotelian Society*, Supp. Vol.
     XXVII (1954), 27-62.   By altogether ignoring the second contribu-
     tion A. J. Ayer contrives to repeat in his *The Problem of Knowledge*
     (Harmondsworth: Penguin, 1956), 170-5, the main mistakes of the
     first.

     Nowadays someone is likely to object that modern physics gives
hospitability to the notion of backwards causation.   Here I can and
will do no more about this than quote yet again from Broad, this
time pillaging his contribution to J. R. Smythies (Ed.) *Science and
ESP* (London: Routledge and Kegan Paul, 1967), p. 195:   "May I add
that it would not be enough to cite eminent physicists who talk as
if they believed this.   What is nonsense if interpreted literally,
is no less nonsense, if so interpreted, when talked by eminent physicists in their professional capacity.   But when a way of talking,
which is nonsensical if interpreted literally, is found to be useful by distinguished scientists in their own sphere, it is reasonable for the layman to assume that it is convenient shorthand for
something which is intelligible but would be very complicated to
state in accurate literal terms."

PROTOSCIENCE, PSEUDOSCIENCE, METAPHORS AND ANIMAL MAGNETISM

Abstract

My aim in this paper is to discuss the concept of 'pseudoscience' in relation to a number of factors which appear to enter into decisions to refer to particular systems of thought as pseudoscientific, and especially to the role of metaphoric thinking in the protoscience/ pseudoscience distinction. Some reference will be made to the use of metaphors in behavioral theory and to a particular example of a pseudoscience, Animal Magnetism.

I.

'Pseudoscience' is a complex concept which, like many complex concepts, may appear to be deceptively simple. Most individuals can name particular pseudosciences: astrology, alchemy, phrenology and animal magnetism are among the better known examples. Most individuals can also name corresponding, closely related areas of legitimate scientific research: astronomy, chemistry, localization of brain function, and hypnosis and suggestibility. There would likely be general agreement that the former are examples of pseudosciences, the latter examples of real sciences. What most individuals might not agree upon is why we make the distinction and what the factors are upon which the distinction is made.

Sometimes a complex concept can best be understood by building it up one piece at a time. The first thing that a pseudoscience must be is a non-science. It would seem axiomatic that a real science cannot be a pseudoscience. So part, but only part, of deciding what we mean by 'pseudoscience' will entail deciding what we mean by 'science.' The second most obvious criterion for a pseudoscience must be that it lays claim to the status of science. There is not much point in calling English literature, art appreciation or theology 'pseudosciences,' even when we can obtain unanimous agreement that they are not sciences, if they make no claim to be sciences.

It is important to recognize here that the *major* distinction which we should make is between 'science' and 'non-science,' *not* between

'science' and 'pseudoscience.' It is to the science/non-science distinction that the traditional demarcation criteria are aimed.[1] Pseudo-
science is one type of non-science, but it does not comprise the whole
of non-science. Neither does "correct" science, which is what is commonly meant by the term 'science' when it is used in contrast to 'pseudoscience,' comprise the whole of science. To neglect a careful delin-
eation of all of the components both of science and of non-science is
to run the risk of seeing the fundamental demarcation as being that
between science and pseudoscience, including in the latter category
such subject matter as "incorrect" science and protoscience which, as
I hope to argue, really do not belong there.

Omitting from consideration traditional ethical, metaphysical,
theological, aesthetic, etc. systems which have commonly neither been
considered to be scientific nor claimed such status, it would appear
that a system which claims to be scientific may meet either one of two
fates. First, it may be shown formally, which has traditionally meant
logically, to be incompatible with criteria generally accepted, at the
time, as being necessary for demarcation as a science. I am not particularly interested in this paper in what these criteria are or
whether, and to what extent, they may vary from one time to another.
Second, it may be shown to meet such criteria, e.g., to be testable,
in principle, through sensory experience. In the former case, it becomes a non-science in terms of the criteria applied, in the latter
case it becomes a science. If the system is categorized as a science,
the possible tests (validation, falsification, or whatever is deemed
appropriate) may actually be carried out. Depending upon the outcome
of the tests, the system might be characterized as either "correct" or
"incorrect" science.[2]

An interesting question here is whether all of the systems of
thought which fall into the first (non-science) category would be considered to be pseudosciences. Certainly, if we define all theories not
deemed 'scientific' by our criteria to be 'non-sciences,' and all 'non-
sciences' which make claims to being 'sciences' to be 'pseudosciences,'
then it follows that all theories falling in the first category are
'pseudosciences.' But this is not what we do in practice. In practice, we would commonly differentiate, within the first category, between those theories which are potentially scientific, but inadequate-
ly formulated, and those theories which we consider to be pseudoscientific.[3]

It would *not* appear to be the case that the logical criteria used
to make the science/non-science distinction will serve to allow us to
make the protoscience/pseudoscience distinction since both of the latter, on the basis of the logical demarcation criteria, are non-sciences. Thus it seems to be the case that, in practice, whatever demarcation criteria are accepted for differentiating between science and
non-science, these criteria may be necessary, but not sufficient, for
differentiating between science and pseudoscience. The same problem
would seem to exist for any number of demarcation criteria. That is,
the simple demarcation of science from non-science, which must precede
the science/pseudoscience demarcation, would never, in itself, be sufficient to differentiate between science and pseudoscience. The reason for this is that within the category of non-science we should always find it useful to make the further distinction between potential
science or naive science or protoscience on the one hand, and pseudo-
science on the other hand. This distinction is always useful since it
keeps us from spending time attempting to develop and refine a system
of thought which we believe to be incapable of becoming a true science.

Further, this distinction is, I shall argue, a matter of judgment
on the part of the scientist and, in practice, eventuates in three possible outcomes. First, the scientist may decide that the non-science-
claiming-to-be-a-science appears to be reformulatable in such a way as
to allow it to meet the demarcation criteria for science. This would
be the potential or protoscience. Second, the scientist may decide
that any reformulation of the system in question that would allow it
to meet the demarcation criteria would be so extensive as to change
completely the basic tenets of the system. That is, the reformulated
system would be so different as to make it meaningless to call it the
'same' system. This would be one way to establish the original system
as a pseudoscience. Third, the scientist may decide that the system
in question could be reformulated so as to meet the demarcation criteria, but it would be meaningless to do so. Take, as an example, the
case of astrology in relation to the single criterion of falsifiability. Astrology could be made to meet the demarcation criterion easily
enough by formulating specific, testable predictions relating the
state descriptions of the heavenly bodies to measurable behavioral
characteristics. How these predictions were formulated should not matter, for example, to Popper since he is not concerned with the origins
of the system to be tested. But, in practice, the attempt to provide

such a formulation is considered to be pointless since we would have
no reason to assign the particular behavioral characteristics to the
state descriptions in anything other than a random manner. In lieu of
an overriding explanation of why star positions at birth should be related to behavior in later life, we should have little hope that the
testable 'science' of astrology which we produced would turn out to be
anything other than an "incorrect" science. Thus, in practice, it
might be deemed possible, but not particularly useful, to reformulate
pseudosciences in acceptable scientific form. In this particular case,
the distinction between pseudosciences and protosciences seems to be
one, in reality, between pseudosciences and potentially "correct" sciences. And this is a factor which we apparently judge without relying
on exhaustive empirical tests. A scientist who undertook a series of
experiments in which he randomly assigned observed behavioral characteristics to star patterns in the hope of finding correlations between
the two would, at the very least, be deemed by the scientific community to be eccentric and exhibiting questionable judgment. So a pseudo-
science is not only a non-science claiming to be a science, but also
it is judged either not to be amenable to reformulation or, if reformulated, not to be a potentially "correct" science. That is, we would
have good reason to believe (or no reason not to believe) that, if it
was reformulated to meet the demarcation criteria, it would be falsified by experience. The analysis thus far might be diagrammed as in
Figure I.

It should be made explicit that to identify the science/non-sci-
ence distinction as being 'logical' is not meant to characterize the
process used in arriving at the demarcation criteria, a process which
may or may not be logical in itself, but rather the process of arriving
at a decision once the criteria have been accepted. Similarly, the
empirical distinction between "correct" science and "incorrect" science
will involve factors, such as the decision of what will count as evidence, which are not in themselves arrived at empirically. But once
those factors have been agreed upon, the process then becomes empirical. These non-rational and non-empirical components of the science/
non-science and "correct" science/"incorrect" science distinctions
are not unimportant. They testify to the complexity of the scientific
process. But they are not the focus of my attention here.

To say the same thing in a different way, both the science/non-
science and the "correct" science/"incorrect" science distinction would

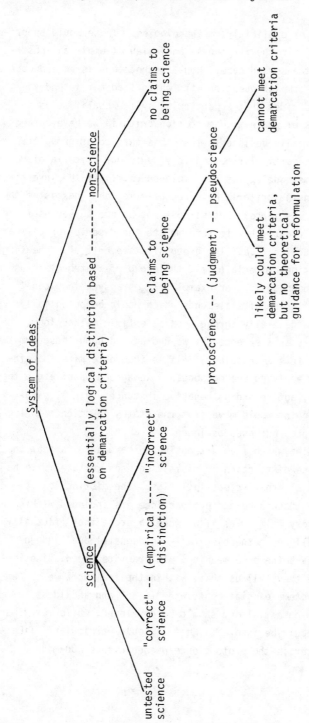

Figure I - Analysis of Science/Non-science Demarcation

include, implicitly or explicitly, methodologies. These would be primarily rules of logic in the first case and rules of empirical observation and test in the second case. How a methodology is agreed upon
is an important and complex question with which I do not intend to
deal here. Having decided that a particular system qualifies as a
science on the basis of whatever demarcation criteria we have agreed
upon, what we have really decided is that this particular group of
statements plus this particular methodology meets our criteria of demarcation. If the methodology is not employed (i.e., if the investigator "cheats") then the criteria have not been met and the system no
longer qualifies as a science. But it does not become a pseudoscience
since we know quite clearly how to reformulate the description of what
is being done in practice (i.e., the group of statements plus cheating)
to meet the demarcation criteria (i.e., the group of statements plus
the agreed upon methodology). For example, *if* parapsychology consisted of a coherent group of statements which would be potentially
testable within some generally agreed upon scientific methodology,
then it would qualify as a science (if we had agreed upon these demarcation criteria and this methodology). If we then determine that investigators are not applying the methodology we agreed upon (e.g., they
are falsifying data, not recording negative instances, etc.), parapsychology at that point would move from the status of science to the
status of protoscience, not pseudoscience.

The general point which I am attempting to make here is that the
science/pseudoscience distinction is misleading when we take it to be
a dichotomy without alternative possibilities. The distinction is
blurred if "incorrect" science and protoscience are included within the
pseudoscience category. It is as if we were to argue that 'factually
incorrect,' 'logically or methodologically inadequate' and 'judged
irredeemable' all have the same meaning. In practice the distinctions
are important since they tell us that, within the framework we are employing, certain systems of statements are "false" and should be rejected, others have potential and should be developed, while yet others
are hopeless and should be ignored. What I should particularly like to
look at in this paper is the protoscience/pseudoscience judgment.

II.

Let's leave the science/pseudoscience problem for the moment and turn to metaphoric thought. I shall use the term 'metaphoric thought' or even 'metaphor' to cover the broad range of cognitive processes involved in the use of analogies, metaphors and models. When it is necessary to make distinctions between these three forms of language, I shall say so and use the three terms individually.

Max Black has observed that,

> To draw attention to a philosopher's metaphors is to belittle him - like praising a logician for his beautiful handwriting. Addiction to metaphor is held to be illicit, on the principle that whereof one can speak only metaphorically, thereof one ought not to speak at all. Yet the nature of the offence is unclear.[4]

Recently, interest in metaphoric thinking has increased markedly as it has come to be realized that this particular cognitive process is something more than merely a clever perversion. Taken broadly, metaphoric thought is a common way of dealing with the world. One psychologist has written,

> Reasoning by analogy is pervasive in everyday experience. We reason analogically whenever we make a decision about something new in our experience by drawing a parallel to something old. When we buy a new goldfish because we liked our old one, or when we listen to a friend's advice because it was correct once before, we are reasoning analogically.[5]

Not only has metaphoric thought come to be seen as common, it has also been argued to be unavoidable. Thus, Morse Peckham has commented,

> ...metaphor is not only a normal semantic mode but a mode essential for the existence and above all the extension of the semantic functions of language. It is the only way we have for saying something new.[6]

Further, the use of metaphoric thought, traditionally viewed as essentially poetic, has been recognized as a necessary component not only of everyday cognitive processes but, more significantly, of scientific thought as well. J. Robert Oppenheimer has argued,

> ...analogy is indeed an indispensable and inevitable tool for scientific progress. ... We cannot, coming into something new, deal with it except on the basis of the familiar and the old-fashioned. ... We cannot learn to be surprised or astonished at something unless we have a view of how it ought to be; and that view is almost certainly an analogy. We cannot know that we have made a mistake unless we can make a mistake; and

> our mistake is almost always in the form of an analo-
> gy to some other piece of experience.[7]

Oppenheimer's last statement is particularly interesting since it implies that the falsifiability criterion is grounded on analogy.

All three of these statements stress the use of metaphoric thought
in dealing with the novel, the new. The use of metaphor is commonly
seen as making understandable the new and largely unintelligible, by
seeing it as something old and familiar. This is the essential function of analogies, metaphors and models. They are used this way both
in our own thinking to make novel concepts more understandable to ourselves, and also in our discourse with other people to help us explain
to them our unfamiliar views. It would appear that if we were to re-
frain from speaking except when we could speak non-metaphorically, we
should be largely silent as we sat in our caves chipping out our stone
hand axes.

To say that the principal function of metaphor is to aid our understanding is to present a non-traditional view of metaphor. Metaphors are usually thought of as literary devices, more concerned with
style than with cognitive content. When we use metaphors for stylistic purposes we generally have what Max Black has called "substitution-
metaphors" or "comparison-metaphors." In these cases, it is normally
possible to replace the metaphor with a literal statement explaining
exactly what the metaphor is intended to convey, without any loss in
cognitive content. Thus, to say that, "the lecturer waded through his
students' term papers" is to make a statement in which the metaphoric
term 'waded' can be replaced by a literal description of the lecturer's
determined and resolute behavior, his resistance to boredom, etc.,
without losing any meaning in the act of substitution. But Black is
more interested in what he refers to as "interaction-metaphors." In
these cases the two portions of the metaphor, which he calls the principal subject and the subsidiary subject, interact on each other and a
simple substitution of a literal statement is not possible without loss
of cognitive content. The point of such a metaphor is understanding,
not style.[8] Thus, to say, "men are becoming machines" is to make a
very complex statement which implies a good deal about the dehumaniza-
tion of individuals, the increasingly "human" aspects of machines, and
the effects of a particular type of social/economic system, many of the
implications of which would be lost in any attempt to replace the subsidiary subject, 'machines,' with a literal statement.

The process of gaining understanding from metaphors can be a particularly complex one. For example, Black argues,

> There are indefinitely many contexts (including nearly
> all the interesting ones) where the meaning of a meta-
> phorical expression has to be reconstructed from the
> speaker's intentions (and other clues) because the
> broad rules of standard usage are too general to sup-
> ply the information needed.[9]

When metaphors are used in a creative way to aid understanding it is
commonly the case that something is known about both the principal
subject and the subsidiary subject of the metaphor. Thus, to say that
the central nervous system is a telephone switchboard can be useful if
we know something, but relatively little, about central nervous systems, and relatively much about telephone switchboards. The problem,
of course, lies in determining in what ways it is useful to think of
nervous systems as being telephone switchboards and in what ways it is
not. Note, not how they are literally similar, but how it pays to
think of them as similar. Again, Max Black has commented,

> We need the metaphors in just the cases when there
> can be no question as yet of the precision of sci-
> entific statement. Metaphorical statement is not
> a substitute for a formal comparison or any other
> kind of literal statement, but has its own dis-
> tinctive capacities and achievements. ...It
> could be more illuminating in some of these cases
> to say that the metaphor creates the similarity
> than to say that it formulates some similarity
> antecedently existing.[10]

We never know, in practice, to what extent we are creating the similarity, just how similar the novel system is to the familiar system,
how appropriate is the metaphor. A large part of the metaphoric cognitive processes then becomes a testing out of the metaphor to see how
far we can use it before it begins to use us. And this is, of course,
the danger in metaphoric thought, the danger of mistaking the mask for
the face, as Colin Murray Turbayne has put it. To think of man as a
machine or the brain as a computer can be extremely useful when we
know relatively little of men and brains and relatively much of machines and computers. But if we are carried away by our metaphors to
the point where we attempt to lubricate men and expect to find that
brains have keyboards, then we are mistaking the mask for the face.
In practice, it is often not clear at what point the metaphor has
stopped aiding our understanding and has begun to pervert it. Precisely because half of the metaphor is relatively unfamiliar, at least
in this particular context, we may not know enough about it to realize
that we have stepped over the limit (to speak metaphorically). Even
when we should know better, the fact that metaphors emphasize certain

aspects of the systems they relate may lead us astray as, for example, where the logical, data processing characteristics of the computer lead us to ignore the non-rational, emotive characteristics of brains.

Black has summarized his views on interaction metaphors in seven claims to which he says the interaction view is committed. They are,

> (1) A metaphorical statement has two distinct subjects - a "principal" subject and a "subsidiary" one.
> (2) These subjects are often best regarded as "systems of things," rather than "things."
> (3) The metaphor works by applying to the principal subject a system of "associated implications" characteristic of the subsidiary subject.
> (4) These implications usually consist of "commonplaces" about the subsidiary subject, but may, in suitable cases, consist of deviant implications established *ad hoc* by the writer.
> (5) The metaphor selects, emphasizes, suppresses, and organizes features of the principal subject by implying statements about it that normally apply to the subsidiary subject.
> (6) This involves shifts in meaning of words belonging to the same family or system as the metaphorical expression; and some of these shifts, though not all, may be metaphorical transfers. (The subordinate metaphors are, however, to be read less "emphatically.")
> (7) There is, in general, no simple "ground" for the necessary shifts of meaning - no blanket reason why some metaphors work and others fail.[11]

But the real crux of the interaction view of metaphors lies in the cognitive value, the aid to understanding, of the metaphors and the fact that they are irreplaceable, without cognitive loss, by a substituted set of literal statements. Black is quite clear on this point when he avers that,

> Metaphorical thought is a distinctive mode of achieving insight, not to be construed as an ornamental substitute for plain thought.[12]

He is referring here to the interesting and non-stylistic uses of metaphor that help us to achieve understanding.

> For substitution-metaphors and comparison-metaphors can be replaced by literal translations (with possible exception for the case of catachresis) - by sacrificing some of the charm, vivacity, or wit of the original, but with no loss of *cognitive* content. But "interaction-metaphors" are not expendable. Their mode of operation requires the reader to use a system of implications (a system of "commonplaces" - or a special system established for the purpose in hand) as a means for selecting, emphasizing, and organizing relations in a different field. This use of a "subsidiary subject" to foster insight into a "principal subject" is a distinctive intellectual operation (though one

> familiar enough through our experiences of learning
> anything whatever), demanding simultaneous awareness
> of both subjects but not reducible to any comparison
> between the two.

> ...One of the points I most wish to stress is that
> the loss in such cases [literal translation] is a
> loss in cognitive content; the relevant weakness
> of the literal paraphrase is not that it may be
> tiresomely prolix or boringly explicit (or defi-
> cient in qualities of style); it fails to be a
> translation because it fails to give the insight
> that the metaphor did.[13]

Thus, metaphors have real cognitive value, they aid insight and understanding, and when they serve that function they cannot be replaced by
the precise, formal, literal, deductive statements which we traditionally have assumed to lie behind merely stylistic metaphoric thought.

Where the cognitive value of metaphoric thought has been recognized, it has commonly been viewed as a heuristic device, not as a form
of argument or explanation. That is one reason why pointing out a
philosopher's metaphors is belittling. It is to remind him that his
arguments are not logical. Thus, metaphoric thought has traditionally
been limited, by philosophers of science and by some scientists when
talking about science, to the processes of discovery rather than the
processes of justification and explanation. And yet, if we deal with
philosophic and scientific practice rather than theory, we quickly come
to realize just how pervasive metaphoric thought is. Indeed, if one
thinks of the application of mathematics in science as an example of
metaphoric thought, of substituting a familiar set of relationships for
an unfamiliar set, then we get a clearer picture of the universality
of metaphor in practice, as well as its usefulness. We also come to
see that the logical gap in the use of metaphor is not within the systems of familiar and unfamiliar concepts, but in the application of
the one to the other.

The wider recognition and use of metaphor in the philosophy of
science has been advocated by Mary Hesse.

> The thesis of this paper is that the deductive model
> of scientific explanation should be modified and sup-
> plemented by a view of theoretical explanation as
> metaphoric redescription of the domain of the explanan-
> dum.[14]

Hesse has relied heavily on a modified version of Black's interaction
view of metaphors. In doing so, she has replaced Black's terms 'principal subject' and 'subsidiary subject' with 'primary system' and
'secondary system.'

> In a scientific theory the primary system is the
> domain of the explanandum, describable in obser-
> vation language; the secondary is the system, des-
> cribed either in observation language or the lan-
> guage of a familiar theory, from which the model
> is taken.[15]

Thus, Hesse's argument explicitly recognizes the possibility of a non-
deductive component in the processes of justification and explanation
as well as in the processes of discovery.

Hesse, like Black, is concerned with meaning and understanding in
science, not simply description. She notes that,

> To understand the meaning of a descriptive ex-
> pression is not only to be able to recognize its
> referent, or even to use the words in the expres-
> sion correctly, but also to call to mind the ideas,
> both linguistic and empirical, that are commonly
> held to be associated with the referent in the
> given language community. Thus a shift of meaning
> may result from a change in the set of associated
> ideas as well as in change of reference or use.[16]

Something like a change in the set of associated ideas may well be effected by factors which lie outside the formal scientific systems.
Therefore, meaning in science is clearly seen to be a matter which has
its external, as well as its internal, referents. Metaphors are not
right or wrong, they are more or less useful, depending on how well
they provide us with understanding, and, perhaps, stimulate our imagi-
nations. It is no more correct to say that "my love is like a red, red
rose" than it is to say that "my love is like six pounds of chopped
liver," or to say that "brains are computers" than it is to say that
"brains are bicycles." Literally, my love is neither a red, red rose
nor chopped liver, the brain is neither a computer nor a bicycle.
Literally, the mind is not a little man in the brain, the nervous system is not a hydraulic system, the eye is not a camera, learning is not
wearing paths, the brain is not a clock, understanding is not seeing.
These are all psychological metaphors which have been both useful and
misleading, but never literally correct. Hesse has noted,

> For a conjunction of terms drawn from the primary
> and secondary systems to constitute a metaphor it
> is necessary that there should be patent falsehood
> or even absurdity in taking the conjunction liter-
> ally.[17]

Literally, things are what they are, and nothing else. Whenever we
speak of them as being something they are not, we speak metaphorically
and, hence, untruthfully (in a literal sense). In a literal sense,
whenever we speak metaphorically we speak falsely, but we do not

necessarily speak uselessly. Thus, it can be useful, in terms of our understanding, to speak untruthfully, providing that we choose our metaphors well. In many ways, choosing our metaphors well is the crux of the problem.

The use of metaphoric thinking in the history of science has been well documented.[18] Here, as in philosophy, the use of metaphor has been formally derided in theory at the same time that it has been extensively employed in practice. Among the better known cases of the use of metaphoric thinking in science have been Galileo's use of analogical argument, based on his observation of Jupiter and the Medicean planets, to show that the earth is a planet; Newton's metaphor of "sociabilities" in discussing the attractive and repulsive properties of magnets; Darwin's use of the tree model in his thinking about evolution.[19] Hesse has commented on the approach taken by philosophers of science to metaphor:

> Acceptance of the view that metaphors are meant to
> be intelligible implies rejection of all views that
> make metaphor a wholly noncognitive, subjective,
> emotive, or stylistic use of language. There are
> exactly parallel views of scientific models that
> have been held by many contemporary philosophers
> of science, namely, that models are purely subjec-
> tive, psychological, and adopted by individuals for
> private heuristic purposes. But this is wholly to
> misdescribe their function in science. Models, like
> metaphors, are intended to communicate. If some
> theorist develops a theory in terms of a model, he
> does not regard it as a private language, but pre-
> sents it as an ingredient of his theory. Neither
> can he, nor need he, make literally explicit all
> the associations of the model he is exploiting;
> other workers in the field "catch on" to its intend-
> ed implications - indeed, they sometimes find the
> theory unsatisfactory just because some implications
> the model's originator did not investigate, or even
> think of, turn out to be empirically false. None of
> this would be possible unless use of the model were
> intersubjective, part of the commonly understood
> theoretical language of the individual theorist.[20]

The choice of an appropriate metaphor can not only help or hinder communication and understanding, but it can also direct research and facilitate acceptance or rejection of a particular theory. That is, in practice, metaphoric thinking which is admittedly non-logical in the sense of non-deductive, can play a role which is not merely heuristic but which is also a factor in judgments made by the individual scientist and the scientific community. I have already argued that the distinction between pseudoscience and protoscience is a matter of

judgment, a judgment made not wholly on logical or empirical grounds.
Metaphoric thought plays a role in that judgment.

III.

It may be useful, at this point, to introduce into our discussion
a specific example. In the late eighteenth century the Austrian physician, Franz Anton Mesmer, claimed to have discovered that "nature af-
fords a universal means of healing and preserving men." The technique,
animal magnetism or mesmerism, involved the transference of magnetic
force or fluid from one individual to another with a healing occurring
in the receiver of the force. Mesmer's system was ridiculed, denounced
and investigated, usually in that order, by scientists and physicians
even while it was being patronized and praised by a number of patients.

Mesmer's ideas and his purported phenomena had their roots in an
intellectual tradition known as hermeticism. Named for a fictitious
Egyptian author, Hermes Trismegistus, the tradition was a varied one
encompassing astrology, alchemy, early iatrochemistry, and natural
magic. Among its basic beliefs was that of man as a microcosm reflecting the macrocosmic universe, and the causal power of likenesses or
'sympathies.' The hermetic tradition in the sixteenth and early seventeenth centuries was, like the mechanical tradition, an opponent of
Aristotelian scholasticism and one of the contenders as a new world
view. But by the middle of the seventeenth century, the mechanical
tradition had won out and hermeticism was well on its way to disrepute.
Its influence lingered and can be found in the writings of such scientific luminaries as Galileo, Bacon, Gilbert, Newton and Boyle. By
Mesmer's time the hermetic tradition was clearly viewed as non-science,
and usually as non-sense.

In 1766 Mesmer had published a thesis in which he described influences which he felt the planets have on human bodies. Utilizing
the metaphor of the Newtonian concept of universal attraction or gravitation, Mesmer argued that there was a direct action of the heavenly
bodies on earthly, animate bodies and particularly on the nervous system. The effect of gravitation in causing the ebb and flow of the
tides led him to argue, by analogy, that there was a similar ebb and
flow in the human body. The term 'animal magnetism' was coined to
indicate "the property of the animal body that renders it liable to
the action of heavenly bodies and of the earth."[21] Mesmer argued that

this was a completely natural phenomenon and that to guide the flow of
magnetic force to specific portions of the body was to work with nature, a process which should have salutary effects.

At the time there were a number of natural phenomena whose existence was accepted and whose general properties were known, but which
were not understood. Among these phenomena were gravitation, magnetism and electricity. Reputable scientists had made attempts to relate gravitation to magnetism and even to suggest that the two were
the same phenomenon. In Mesmer's earlier writings, when his interest
was directed to the attractions exerted by planetary bodies on animal
bodies, the gravitational metaphor loomed large, with references to
magnetism which suggest that gravitation and magnetism might be identical. In later writings the gravitation metaphor disappeared almost
entirely as Mesmer's interests shifted to influences between animal
bodies, and the magnetism metaphor was used almost exclusively, with
occasional suggestions of an electricity metaphor. Let us for the
moment take seriously Black's comment (cited on page 85) that "the
meaning of a metaphorical expression has to be reconstructed from the
speaker's intentions." If one views Mesmer as most authors have, as a
conscious fraud and charlatan, then the metaphors may be interpreted
as a crass attempt to give medical humbug some scientific respectability. Alternatively, if Mesmer realized that his cures were psychological, not physical, but wanted to give his patients confidence and
reassurance in physical terms he felt they would accept, then his metaphors may be interpreted as part of the therapeutic procedure. As yet
another possibility, if Mesmer genuinely believed in the existence of
the magnetic fluid and its healing properties, and if he further believed that they were all natural, physical phenomena, then his metaphors may have been serving the purpose that they appear to serve for
all scientists, that of aiding understanding and communication by
placing the unfamiliar within a familiar context.

We cannot, with any assurance, read Mesmer's intentions at this
late date. However, to assume without further evidence that he must
have been a charlatan because *our* standards of scientific knowledge
make animal magnetism unbelievable is, of course, to condemn Galileo,
Bacon, Gilbert, Newton and Boyle as charlatans for some of their beliefs. We have come face-to-face with the problem of a demarcation
criterion that will rule out what we wish it to rule out without ruling out that which we do not wish it to rule out. Other evidence

would not particularly support the contention that Mesmer was a conscious fraud. Mesmer's advocacy of animal magnetism can hardly be
characterized as the ploy of an ignorant confidence trickster looking
to improve his fortunes by way of the fast buck. He was well educated,
having studied music, philosophy, theology and law before taking a
doctorate in medicine at the University of Vienna. The medical faculty
at Vienna was one of the best in Europe, under the direct control and
patronage of Maria Theresa and under the direction first of Gerhard van
Swieten and then Anton de Haen, both students of Hermann Boerhaave.
Mesmer's dissertation topic, *The Influence of the Planets on the Human
Body*, was controversial and hardly the kind of subject that would have
been chosen by a sharpster eager to please the medical establishment.
The hermetic tradition was widespread in German speaking Europe and
societies such as the Rosicrucians were popular. But while Mesmer's
topic may have been in the hermetic tradition, his handling of it was
not, nor is there any reason to believe that the Faculty of Medicine
would have tolerated a thesis in mysticism. Mesmer argued for a material, not a mystical, relationship in terms of a "universal fluid" that
was an acceptable part of eighteenth century mechanical science. In
1768 Mesmer married into a wealthy aristocratic family, the wedding
ceremony being performed by the Archbishop of Vienna. Mesmer and his
wife settled into a large mansion in a fashionable section of Vienna
and moved in the social world of the Viennese aristocracy. Mesmer had
no need of additional money and socially there was nowhere to climb.
His financial security and his love of music led him to become the pa-
tron of a struggling young musician, Wolfgang Amadeus Mozart. It was a
good life. Beginning in approximately 1773, Mesmer's stubborn and passionate advocacy of animal magnetism, far from being a way *up* the financial and social ladder, turned out in fact to be a way *down*. It was
an embarrassment to his colleagues, his family and his social acquaintances. The point is that while we may not be able to state with certainty what Mesmer's intentions were in using mechanical metaphors to
explain hermetic subject matter, it is virtually certain that his intention was *not* out-and-out fraud.

Nor, it would seem, were his metaphors directed at his patients
who appear to have been well pleased with the "cures" they experienced,
whatever the explanation. Rather, his attempted expositions of his
views and his appeals for a chance to demonstrate and explain his phenomena were directed to the scientific academies of Europe, most of

which ignored him. His metaphors were meant to be explanatory and com-
municative, part of the normal processes of scientific discourse. Moreover, it is clear that Mesmer knew that his references to magnetism
were metaphoric, while at least some of his critics took them to be
literal.

The phenomena which the magnetic fluid transfer was designed to
explain were primarily convulsions and healing. Patients being treated
by Mesmer would sometimes, but not always, pass from a state of drowsi-
ness and lethargy into a series of convulsions referred to as the
"crisis." The convulsions were thought to be particularly efficacious
in the process of healing which then was supposed to follow. Claims of
cures were made for ailments ranging from rheumatism to blindness, head-
aches to cancer. Moreover, the claims were attested to by a large number of testimonials from patients, their relatives, impartial observers
and, occasionally, clergy and physicians. There were also a number of
secondary phenomena reported by mesmerists but not claimed by Mesmer to
be consistent or necessary accompaniments of his treatment. They included feelings of "prickling," pain and heat, hiccups, and uncontrol-
able weeping and laughter.

In 1779, Mesmer published a brief "Dissertation on the Discovery
of Animal Magnetism." By this time he was already considered by most
of the medical and scientific community to be a charlatan, and the
Dissertation is, in effect, his side of the controversy. In addition
to a historical summary of the "facts," Mesmer presented twenty-seven
propositions which can be taken as a definitive statement of his system of animal magnetism. There seems to have been an evolution in
Mesmer's thought beginning with the gravitational/magnetic influence of
the planets and progressing through a stage in which he believed that
actual magnets had to be used, by placing them on appropriate parts of
the patient's body, to effect a cure. Finally, he came to realize that
the magnets were not necessary, that cures could be obtained simply by
touching or stroking the patient, and even by pointing from a distance.
But having argued from the beginning that the curative agent was the
magnetic fluid or force which passed to the patient, Mesmer seemed unable to drop that part of his system when magnets and planets were no
longer considered essential. It is at this point that the references
to magnetism became metaphoric. The flow of the magnetic fluid was
then described as occurring between individuals. This clearly could
not be mineral magnetism but, like mineral magnetism, it was a force

which was unobservable except through its effects. It was, therefore, animal magnetism, a metaphor which Mesmer thought would be helpful but which turned out to be confusing.

Mesmer did continue to make use of one non-animate device, the *baquet*. This was a large wooden tub or chest filled with water and containing a number of pieces of metal. Iron rods protruded from the tub and the patients would be seated in a circle surrounding the *baquet*. A cord would be used to join all the patients together (suggesting a "conductor" of some kind) and they would sometimes hold hands. Mesmer and his assistants would pass among the patients touching their bodies with the iron rods protruding from the *baquet*. Following this proce- dure, which was usually accompanied by music, the various behavioral phenomena already described would begin to appear.

The central assumptions of his system are summarized in the first four of Mesmer's twenty-seven propositions. They are,

1. There exists a natural influence between the Heavenly Bodies, the Earth and Animate Bodies.
2. A universally distributed and continuous fluid, which is quite without vacuum and of an incomparably rare- fied nature, and which by its nature is capable of receiving, propagating and communicating all the impressions of movement, is the means of this in- fluence.
3. This reciprocal action is subordinated to mechanical laws that are hitherto unknown.
4. This action results in alternate effects which may be regarded as an Ebb and Flow.[22]

Within the framework of the science of Mesmer's time, none of these pro- positions would be obviously non-scientific or nonsensical. He is say- ing that there is an influence between material bodies which can be discerned by means of its effects. In mechanical terms, forces can only be transmitted between bodies through an intervening medium. This is the rarefied fluid and its existence is meant to be literal. Its action is controlled by mechanical laws (i.e., it is not spiritual), but the laws are not necessarily the mechanical laws already known at the time. Given the knowledge which existed in the science of Mesmer's day, this set of statements could be a description of gravitation, electricity or mineral magnetism. Each of these possible metaphors, and one more, are suggested in Mesmer's twenty-seven propositions. The over-riding metaphor is, of course, magnetism. But Proposition 1 also suggests the gravitational metaphor. Proposition 15 reads, "It is in- tensified and reflected by mirrors, just like light", specifically mak- ing the analogy to light. Proposition 17, "This magnetic property may

be stored up, concentrated and transported," suggests the electricity
metaphor, the properties of the Leyden jar having been known since approximately 1746. These analogies and metaphors are all to natural,
physical phenomena, albeit phenomena not clearly understood or explained in 1779. It is an attempt to place the healing phenomena of the
hermetic tradition within the explanatory framework of the mechanical
tradition. This was Mesmer's intention, and it is within this framework that we must understand his metaphors.

That Mesmer understood the references to magnetic phenomena to be
metaphoric was brought out in two propositions, 9 and 20.

> 9. It is particularly manifest in the human body
>   that the agent has properties similar to those
>   of the magnet; different and opposite poles
>   may likewise be distinguished, which can be
>   changed, communicated, destroyed and strength-
>   ened; even the phenomenon of dipping is ob-
>   served.[23]
> 20. The Magnet, both natural and artificial, together
>   with other substances, is susceptible to Animal
>   Magnetism, and even to the opposing property,
>   without its effect on iron and the needle under-
>   going any alteration in either case; this proves
>   that the principle of Animal Magnetism differs
>   essentially from that of mineral magnetism.[24]

What Mesmer was arguing was that magnets and human bodies have *some*
similar properties, *not* that human bodies are magnets. He is at pains
to distinguish between animal and mineral magnetism, and to note specifically that the effects of animal magnetism on iron and the needle,
through a mineral magnet, are quite different than would be the effects
of mineral magnetism.

Another scientific controversy of Mesmer's time, relating to the
existence of a universal fluid or aether, is suggested by Proposition
14.

> 14. Its action is exerted at a distance, without the
>   aid of an intermediate body.[25]

Newton was uneasy with the idea of action at a distance. He refused to
accept gravitation as a *property* of matter and attempted (unsuccessful-
ly) to provide a mechanical explanation of its effects. The same problem existed for magnetism. Richard Westfall has commented,

> Magnetic attraction, indeed, was a crucial pheno-
> menon to proponents of mechanical philosophies of
> nature - the very epitome of the occult sympathies
> and antipathies with which Renaissance naturalists
> had populated the universe. Unless mechanical
> philosophers could explain it away, the assertion

that nature consists solely of matter in motion
was, without further ado, shown to be false.[26]

Descartes attempted a mechanical explanation of magnetism in terms of
right and left-hand threaded (to account for polarity) screw-shaped
particles of matter. Newton chose to speculate on a system of two distinct odiferous streams of magnetic matter with different "sociabilities" (attractions and repulsions). Mesmer's claim, in Proposition 3,
that unknown mechanical laws controlled the flow of the animal magnetic
fluid was, therefore, not any different from the claims that many respected scientists had made concerning both mineral magnetism and gravitation. But Westfall's comment brings out clearly the importance of
the metaphor for Mesmer. Action at a distance without the mechanical
metaphor is a hermetic solution to a hermetic problem, to be rejected
out of hand in the late eighteenth century. But with the mechanical
metaphor, Mesmer might well claim that his explanatory system was no
worse off than those of the main stream mechanical philosophers in relation to these types of phenomena. Where Mesmer differed from the
mechanical philosophers was in the use of mathematics. Mesmer's was
not a quantitative system nor a system that presented its proposed laws
in the form of mathematical equations. This was a characteristic of
the hermetic tradition which Mesmer retained.

Thus far, there would be little or no reason to characterize Mesmer's system as a pseudoscience within the framework of the science of
his time. On the face of it, he was attempting to explain the observed
phenomena within a scientific, even a mechanical, framework. He was
attempting to provide explanations not in terms of the "sympathies" of
Renaissance natural magic, but in terms of the subtle aethers and mechanical forces of eighteenth century mechanistic science. His speculations seem not to have been essentially different in kind from those
of the accepted scientists of his day. Even the tone of his *Dissertation* was strongly empiricist, another characteristic which appeared, in
strange ways, in the hermetic tradition. He warned against forgoing
observations and building up a rational system with little or no connection to observed facts, indulging in "frivolous speculation." He
continually emphasized that he could *see* the effects on his patients
which he then attributed to animal magnetism.

In 1784 Louis XVI established a Royal Commission to investigate
animal magnetism. The members of the commission were drawn from among
the leading physicians and scientists of the time. Its President was
Benjamin Franklin and among its members were Lavoisier, Guillotin and

Bailly. The commission dealt not with Mesmer, but with Deslon, a physician who was one of Mesmer's followers. The report of the commissioners is particularly interesting in that it provides some insights
into the methods decided upon by the group to test the claims of the
animal magnetists. They began by observing the reactions of patients
during a session held by Deslon making use of the *baquet*. The phenomena they report observing are essentially the same as the phenomena that
Mesmer had reported, including the convulsive 'crisis.' Indeed, nowhere in their report do the commissioners ever question the existence
or the genuineness of the behavioral phenomena, or suggest that the patients do not really experience the sensations which they claim to feel,
or charge that the patients are faking the seizures. To this extent,
the existence of the behavioral phenomena are never in question. This
is an important point since Deslon, as the commissioners observe, argued that the *only* way to demonstrate the existence of the animal magnetic fluid was through its effects on the patients. The commissioners,
as we shall see, accepted many of the effects but not their purported
causes.

One of these effects was, of course, the reputed cures. This, for
Deslon (and, one would assume, for Mesmer), was the heart of the issue.
The evidence here consisted of hundreds of testimonials from former patients who claimed to have been cured by the magnetic procedures. The
commission, not surprisingly, ruled out consideration of these testimonials on the grounds that many illnesses, if left untreated, run their
course and are cured through the unaided action of nature. Since it
would not be possible to determine which of the cures attributed to
animal magnetism were of this kind and which were not, the reputed
cures were not considered to be evidence. This dismissal of the cures
from consideration and the consequent failure of the commissioners to
realize the importance of the phenomenon of functional cures led Frank
Podmore, writing in 1909, to suggest that,

> The 11th of August should be observed as a day of
> humiliation by every learned Society in the civil-
> ized world, for on that date in 1784 a Commission,
> consisting of the most distinguished representatives
> of Science in the most enlightened capital in Europe,
> pronounced the rejection of a pregnant scientific
> discovery - a discovery possibly rivalling in perma-
> nent significance all the contributions to the Physi-
> cal Sciences made by the two most famous members of
> the Commission - Lavoisier and Benjamin Franklin.[27]

What Mesmer considered to be his greatest triumph was the case of
Maria Theresa Paradies, daughter of the empress's private secretary.
This unfortunate girl, an accomplished pianist, awoke one morning at
the age of three with a total and inexplicable blindness. The ailment,
almost surely hysterical in nature, was treated by numerous physicians
for fifteen years without any success. The treatments included bleed-
ing by leeches and the application of hundreds of electric shocks to
her eyes. Finally Mesmer was called in and, on the testimony of the
girl's father, had her seeing within days. One might argue that spontaneous remission of the blindness could have occurred at just this
time, after fifteen years, without Mesmer's help, but this seems unlikely. It should have been a clue to the commissioners that a closer
look at the cures, no matter what their cause, would have been worthwhile. The physicians who had treated the Paradies girl without success over a long period of time, had already declared her to be incur-
able. Therefore, in their opinion, Mesmer could not have effected a
cure. Professor Joseph Barth, an opthalmologist who had not been successful with Maria Theresa, examined her after Mesmer's treatment and
declared her not cured. According to Mesmer, Barth's examination consisted of presenting the girl, who had not seen for fifteen years since
the age of three, with common objects which she was required to name.
When she could not, or was confused, Barth apparently concluded that
she could not see them rather than that he had provided yet another
response to Molyneux's question.

What remained to be tested? The commission's attention turned to
the actual existence of the magnetic fluid or force and the question of
whether the behavioral phenomena could be produced under controlled
conditions. From the beginning, the critique of animal magnetism centered on two, largely distinct, questions. The first involved the validity of the cures and Mesmer's long feud with the medical profession.
The second involved the physical evidence for the existence of the animal magnetic fluid. Mesmer's propositions of 1779 were almost exclusively directed to the latter issue as were his approaches to the various scientific societies, and his appeal at that stage was being made
to the scientists rather than to the physicians who had already dismissed him as a charlatan. However, the scientists to whom he was appealing were part of a different tradition. In the hermetic tradition
it was an accepted principle that objects in the world had two aspects,
a visible physical body and a non-visible 'spirit' or 'essence' which

could be known only through its effects. The scientists of the late
eighteenth century were concerned with explanations in terms of mechanical causation and *consistent* effects under controlled conditions. It
was, therefore, by those rules that Mesmer attempted to play through his
mechanical metaphors. But the metaphors were intended to make the explanations of the phenomena acceptable and to aid discussion. Taken
literally, they suggested to the commission ways in which the very existence of the animal magnetic fluid might be tested. And within their
tradition, establishing the existence of the fluid would not be easy.
The commission commented,

> The most certain method of determining the existence
> of the animal magnetic fluid, would have been to have
> rendered its presence capable of being perceived by
> the senses; but it did not take long to convince the
> commissioners that this fluid was too subtle to be
> subjected to their observation. It is not, like the
> electrical fluid, luminous and visible; its action
> is not, like the attraction of the lodestone, the
> object of one's sight; it has neither taste nor
> smell, its process is silent, and it surrounds you
> or penetrates your frame, without your being in-
> formed of its presence by the sense of touch. If
> therefore it exists in us and around us, it is after
> all perfectly invisible.[28]

If we recall that Mesmer, in his Proposition 20, pointed out that the
effects of animal magnetism were quite different from the effects of
mineral magnetism, specifically in relation to iron and the needle,
then some of the tests conducted by the commissioners seem strangely
irrelevant. For example, they say, in relation to the *baquet*,

> The commissioners in the progress of their examina-
> tion discovered, by means of an electrometer and a
> needle of iron not touched by the lodestone, that
> the bucket contained no substance either electric
> or magnetic; and from that detail that M. d'Eslon
> made to them concerning the interior structure of
> the baquet, they cannot infer any physical agent,
> capable of contributing to the imputed effects of
> the magnetism.[29]

The purported function of the *baquet* was the storage and transmission
of the *animal* magnetic fluid. Since Mesmer had already stated that the
animal magnetic fluid did not cause the same effects on these measuring
instruments as did mineral magnetism, it is difficult to see what conclusions, if any, could be drawn from the commissioners' tests concerning the existence, or non-existence, of the *animal* magnetic fluid. The
commissioners here seem to have taken Mesmer's metaphor to be literal
and tested for the wrong effects. They were being used by the metaphor,

mistaking the mask for the face. Deslon's frustration must have been
acute. The physicians had failed to appreciate the importance of the
phenomenon of psychological cures, the scientists were testing for the
existence of phenomena which Mesmer had specifically denied existed.
Intellectual traditions were clashing and the metaphors were not helping.

The crucial tests, in the minds of the commissioners, were made on
themselves and on a number of individuals of varying ages and social/
educational backgrounds. These tests were all conducted in private,
under controlled conditions. They involved blindfolding patients to
see if they could tell, without seeing the magnetizer, to which parts
of their bodies the animal magnetic fluid was being directed; having
the magnetizer screened from the sight of the patient who did not know
that he/she was being treated; and having the commissioners themselves
treated to see if they noticed any effects. In general, the findings
were that, when the patient could see to which part of his/her body the
magnetizer was directing the magnetic fluid (by pointing or passing his
hands over the patient), then the patient would report feeling prickling sensations, heat or pain, in the appropriate part of the body.
When the patient could *not* see the magnetizer, then the various sensations were reported to be felt randomly in parts of the body other than
those to which the magnetizer was directing the force. When patients
knew they were receiving the attentions of the magnetizer, they would
reach the stage of convulsive crisis in a matter of a few minutes. But
when the magnetizer was screened from patients who did not know he was
there, he could perform the magnetizing procedures for long periods of
time without any apparent effect on the patient. The commissioners reported that they, themselves, felt no sensations while the magnetic
force was being directed at them, other than the usual body sensations
which one perceives when attending to them. Blindfolded patients who
believed in animal magnetism would reach the extreme 'crisis' stage
when told that Deslon was in the room and performing the magnetic procedures, even when he was not.

The conclusions of the commission were clear:

> The experiments which we reported, are uniform in
> their nature, and contribute alike to the same de-
> cision. They authorize us to conclude that the
> imagination is the true cause of the effects at-
> tributed to magnetism.[30]

> This fluid is said to circulate through the human
> body and to be communicated from individual to

> individual. ...This agent, this fluid, has no
> existence. ...The magnetism is no more than an
> old falsehood. The theory is now presented with
> a greater degree of pomp and ceremony, but it is
> not less erroneous.[31]

> It appears by the experiments we have related
> that imagination alone produces the crisis.[32]

> Since the imagination is a sufficient cause, the
> supposition of the magnetic fluid is useless.[33]

Thus, the general conclusion of the commission was that through a series of controlled scientific tests they were able to disprove the crucial assertion of the system of animal magnetism, the existence of a
magnetic fluid or force, and to show that the real cause of the undis-
puted behavioral phenomena was imagination. One member of the commission, Laurent de Jussien, published a dissenting report, and a secret
report purportedly was sent to the king discussing the sexual aspects
of the treatment and cures.

From the report of the commissioners, what can we say about the
status of animal magnetism? Is it "incorrect" science, protoscience or
pseudoscience? The physicians, at least, seem to have decided on the
last of these categories long before the report of the commissioners.
Thus, when Mesmer asked Anton von Stoerck, his former professor, his
friend, and his colleague, head of the Faculty of Medicine in Vienna
and personal physician to the empress, to at least investigate the curative effects of animal magnetism, von Stoerck flatly refused and, instead, requested that Mesmer stop discussing his views publicly. But
the commission seems to have opted for animal magnetism as "incorrect"
science. It is difficult to see how else one can interpret a situation
where a system is subjected to controlled, scientific tests, and found
wanting. Pseudosciences are, as we have seen, non-sciences. Whatever
other characteristics they may have, or whatever demarcation criteria
may have been used, one point would seem clear - that scientific tests
are applicable and can lead to conclusions only in the case of scientific systems. That is, if we accept the testimony of the commissioners, we should also accept Mesmer's twenty-seven propositions as an example of scientific theory. Alternatively, if Mesmer's theory was not
science, or was pseudoscience, then it is difficult to see how it could
be testable by scientific means, and the commissioners' experiments
must have been irrelevant. But it is not at all clear how we can call
mesmerism 'pseudoscience' and at the same time maintain that it was
shown to be invalid by scientific tests. To do so is to confuse

"incorrect" science with pseudoscience, and it is hardly clear that such a confusion would lead to an improvement in our understanding of what does and what does not constitute pseudoscience.

Most demarcation criteria that have been proposed for the science/ non-science distinction have, at the very least, included the possibility of empirical test as a portion of the criteria. By most criteria, therefore, I believe that Mesmer's system was clearly non-science, and that the commissioners were mistaken in treating it as a science by performing controlled empirical tests. Most, if not all, of their tests, like the test for mineral magnetism in the *baquet*, were irrelevant to Mesmer's claims. To show why this was the case, we must go back to Mesmer's propositions again. I shall attempt to show that they were untestable because, first, they were internally inconsistent and, second, they provided explanations for any conceivable outcome of any conceivable empirical test. It may reasonably be argued that it is unfair for us to expect the commissioners to apply our twentieth century demarcation criteria to their eighteenth century problems. However, even if the commissioners did not recognize untestability as a formal demarcation criterion of a non-science, they should at least have recognized untestability as a good reason for not carrying out tests.

As an example of the internal inconsistency, consider propositions 13 and 15:

> 13. Experiments show the passage of a substance
> whose rarefied nature enables it to penetrate
> all bodies without appreciable loss of activity.[34]
> 15. It is intensified and reflected by mirrors, just
> like the light.[35]

Thus, the animal magnetic fluid both penetrates *all* bodies and, at the same time, is reflected by mirrors. But the real untestability of the system appears in Propositions 18 and 19.

> 18. I have said that all animate bodies are not
> equally susceptible; there are some, although
> very few, whose properties are so opposed that
> their very presence destroys all the effects
> of magnetism in other bodies.[36]
> 19. This opposing property also penetrates all bodies;
> it may likewise be communicated, propagated,
> stored, concentrated and transported, reflected
> by mirrors and propagated by sound; this consti-
> tutes not merely the absence of magnetism, but a
> positive opposing property.[37]

We have not only an animal magnetic force, but also an anti-magnetic force which has all of the same properties as the magnetic force, except that it produces the opposite effects. Here we must recall Deslon's comment to the commissioners that the only way he could prove the existence of the magnetic force was through its results. It now becomes clear that while positive results would be taken as evidence for the existence of the animal magnetic fluid, negative results would not be evidence against its existence but, rather, would be evidence for the existence of the anti-magnetic force. This is a classic example of the invention of 'forces,' 'faculties' and 'properties' to account for observed phenomena, each phenomenon having its own unique hypothesized cause. Moreover, Propositions 18 and 19 are inconsistent with Proposition 7 which reads,

> 7. The properties of Matter and the Organic Body
>   depend on this operation. [The ebb and flow
>   of the animal magnetic fluid.]

If the properties of everything depend on the action of the magnetic fluid, what power can anything else, such as the anti-magnetic fluid, have? If the magnetic fluid fills everything without any empty spaces (Proposition 2), where can the anti-magnetic fluid, which also penetrates everywhere, actually exist? This is, in every way, non-sense. The report of the commissioners makes no mention of the anti-magnetic fluid or the logical inconsistencies and, indeed, their procedures and conclusions are only valid if one ignores the existence of the second force. In doing so, of course, it is clear that whatever they might have been testing, it was not Mesmer's system of animal magnetism. That, I would maintain, remains untestable. Moreover, it is difficult to see how a system that is essentially inconsistent and empirically unfalsifiable in the way that Mesmer's system was, could ever be more explicitly formulated to make it acceptable as a scientific system, whether "correct" or "incorrect." It qualifies, therefore, not only as a non-science, but as a real pseudoscience.

One last comment on the report of the commissioners. If we accept their conclusion that their tests proved the non-existence of the animal magnetic fluid, then we might wish to look at their alternative explanation. It was, you will recall, that the behavioral effects were caused by 'imagination,' which they refer to as an already known cause of behavioral phenomena. Concerning imagination, the commissioners say,

> The imagination of sick people has unquestionably
> a very frequent and large share in the cure of
> diseases. With its effect, we are otherwise not
> acquainted except through general experience.
> Though it has not been traced in positive ex-
> periments, it should seem reasonable not to admit
> doubt. It is a well-known adage, that in psychic
> as well as religion, men are served by faith.
> This faith is the product of the imagination. In
> these cases the imagination acts on a gentle means,
> and it acts by diffusing tranquility over the sen-
> ses, by restoring the harmony of the functions, by
> recalling into play, every principle of the frame
> under the genial influence of hope.[39]

Having accepted imagination as the cause of the behavioral phenomena in place of animal magnetism, the commission has now noted that (1) the principles upon which imagination works are largely unknown, (2) the power of imagination is known through its effects, such as healing, and (3) there is no experimental evidence which would establish its existence or the principles of its operation. The evidence put forward in its favor is a "well-known adage" and some speculation on religious faith and hope. Why should the tough minded scientists who rejected animal magnetism and the magnetic fluid have abandoned their scientific criteria and accepted, without evidence or explanation, the causative power of imagination? The answer is probably that imagination, as the commissioners term it, is a 'psychic' phenomenon, not a physical phenomenon and certainly not a mechanical phenomenon. It lies outside of their scientific/mechanistic tradition and so need not meet whatever criteria that tradition lays down for mechanistic, causal explanations. Mesmer had chosen his metaphors so as to place his phenomena within the tradition of the mechanical philosophy. He could, therefore, be judged by the criteria of that tradition. But who knows what imagination is? Who knows what criteria are applicable to imagination as an explanatory device? Alfred Binet and Charles Féré, writing in 1888, comment as follows on the commission's attribution of the observed phenomena to imagination,

> In our day, this would appear to be an insufficient
> explanation. We might as well say that hysteria is
> due to the imagination.[40]

To summarize, Mesmer took real phenomena, functional cures of hysterical disorders, which had traditionally been considered a part of the hermetic tradition of renaissance natural magic and attempted, through the use of metaphors, to make those phenomena acceptable to the dominant mechanical philosophy. The metaphors were taken literally, the

metaphoric explanations of the phenomena were tested within the framework of mechanical causes and effects, and were rejected. An explanation no more substantial than that of the animal magnetic fluid was accepted because it was not seen to fall within the scope of the mechanical philosophy and, hence, was not evaluated with the same criteria.
Mesmer might have been well advised to keep his metaphors to himself.

IV.

In the light of the foregoing discussion, the status of animal
magnetism as a popularly recognized pseudoscience becomes both more
complex and more interesting. It would appear that in the latter part
of the eighteenth century, animal magnetism was close to being all
things to all men. Its proponents saw it as a legitimate science or,
at worst, as a protoscience needing development. The scientists saw
it as an "incorrect" science, a system of ideas that had been subjected
to scientific tests and found to be wrong. The physicians, on the
whole, dismissed it out of hand as medical humbug, an obvious pseudo-
science. We have already looked at the first two of these categoriza-
tions in Mesmer's propositions and the commissioners' report, and turn
now to the medical conclusions.

One historian of hypnotic phenomena has commented,

> To the student of the history and philosophy of
> science, the development of theories about hypno-
> sis is particularly fascinating. Few other psy-
> chological problems provide such clear-cut exam-
> ples of the effect of the *Zeitgeist* on theory and
> practice, of the influence of contemporary scien-
> tific models on the explanation of observed phe-
> nomena, of the interpenetration of ideas developed
> by philosophers, neurologists, psychologists, and
> practitioners of the healing arts. In few other
> areas can one find such strong partisanship for
> theories, producing challenges and counterchallen-
> ges.[41]

One might expect that while the influences of the *Zeitgeist* and views
of contemporary scientific models might be "particularly fascinating"
in relation to hypnotic phenomena, they would not be influences that
are particularly unique to this area of research. In order to understand judgments that a particular system is pseudoscience at any given
moment in time, we must understand not simply the logic of the system,
but also factors relating to the social, intellectual and scientific
thought of the time. This suggests that the pseudoscience/protoscience

distinction (and, incidentally, the pseudoscience/science distinction)
is not absolute, but one that is very much a part of the general pattern of thought which exists at any given time.

The major phenomenon claimed by the animal magnetists, indeed, the
whole point of the magnetic procedures, was healing. The beginnings of
belief in healing through touching are lost in antiquity. Sir James
Frazer, writing in the *Golden Bough*, has commented,

> Royal personages in the Pacific and elsewhere have
> been supposed to live in a sort of atmosphere high-
> ly charged with what we may call spiritual electri-
> city, which, if it blasts all who intrude into its
> charmed circle, has happily also the gift of making
> whole again by a touch. We may conjecture that simi-
> lar views prevailed in ancient times as to the pre-
> decessors of our English monarchs and that accord-
> ingly scrofula receives its name of the King's Evil
> from the belief that it was caused as well as cured
> by contact with a king.[42]

What Frazer was referring to here was the ancient practice of the kings
of both England and France of "touching for scrofula." It was believed
that the king's touch could cure this particular disorder, and earlier,
that the royal touch was a general cure. Marc Bloch has traced the
history of this belief and practice to medieval times and has shown its
relationship to belief in the monarch as the anointed of God and in
the sacrament, at the time of coronation, of unction. The healing powers began to fall into disrepute during the reign of Louis XV and the
ceremony had to be cancelled on a number of occasions when the king's
confessor refused to allow him to partake of the sacraments. These,
Bloch reports, were occasions of great scandal in Paris and helped to
further weaken the prestige of the monarchy. Even earlier, Voltaire
had directed his potent ridicule at this practice, noting that one of
Louis XIV's mistresses had died of scrofula although she had certainly
been well touched by the king. Thus, the power to heal by touching was
a factor in the decline of a political/social system, and not an isolated medical or scientific phenomenon. Although it is not known for
sure when Louis XVI last touched for scrofula, Bloch reports that the
size of the crowds who appeared for the ceremonies diminished markedly
during the eighteenth century, indicating a decline in belief in the
practice. The last of these ceremonies must have been held at approximately the same time that Mesmer published his propositions and Louis
XVI set up his investigative commission. It would seem reasonable to
assume that general skepticism about the monarch's ability to heal by

touching, an ancient rite closely tied to both government and church, might have been translated into a skeptical attitude towards *any* ritual where touching was supposed to lead to cures. Louis, of course, might be expected to have been uneasy about any system which suggested that non-royal personages could heal by their touch.

Clearly, also, physicians would have had good reason to oppose Mesmer's suggestion that this healing ability was virtually universal. The opposition from the medical establishment came early, long before any adequate logical or empirical tests of Mesmer's system were carried out. Individual physicians, such as Deslon, who supported Mesmer had their professional recognition and privileges withdrawn. The conclusion reached by the medical profession, that animal magnetism was a pseudoscience, seems not to have been based on any rational analysis of Mesmer's claims, except in as much as it is always rational to protect one's own financial interests. Looking back on eighteenth century medicine from our perspective, a good deal of charlatanism, humbug and pseudoscience is perceivable. What made animal magnetism a particular target must, therefore, have been related to a number of socially conditioned opinions and beliefs of the time.

The royal tradition of touching for scrofula was, at least in part, a spiritual tradition. The power of the monarch to cure was a sign from the deity of confirmation of the temporal power to rule. But there was also a tradition of curing by magnetism which was not particularly related to theological claims. The commissioners, in their report to the king, mention three of Mesmer's predecessors.

> But the imagination is that active and terrible
> power, by which we are accomplishing the astonish-
> ing effects that have excited so much attention to
> the public process. The effects strike all the
> world, the cause is enveloped in the shades of
> obscurity. When we consider that these effects
> seduced in former times men, venerable for their
> merit, their illumination and even their genius,
> Paracelsus, Van Helmont and Kircher, we cease to
> be astonished that persons of the present day,
> learned and well-informed, that even a great num-
> ber of physicians have been dupes of this system.[43]

The reference to Paracelsus, Van Helmont and Kircher, while recognizing in passing their abilities, is clearly disparaging in reference to their views on healing. Eighteenth century readers would have known these individuals as a part of the discredited hermetic tradition, practitioners of alchemy and astrology as well as iatrochemistry.

Paracelsus (1493-1541), whose real name was Theophrastus Bombas-
tus von Hohenheim, was a sixteenth century physician/chemist/alchemist,
who has been referred to by one historian of physiology as, "this pic-
turesque man, whose name has become a by-word for fantastic thought
and even for charlatanry."[44] He was author of the first work on the
occupational diseases of miners, and argued that nature consists of
visible matter and invisible forces, a distinction not limited, in the
history of science, to charlatans. Through his alchemical studies he
became interested in the curative effects of various ores and metals.
This was an approach very strongly rejected by the established medical
profession of his time, although later scientists such as William Gilbert (in *De Magnete*) gave serious discussion to the matter. As Sir
Michael Foster has observed,

> It may be noted that those who were engaged in the
> search for the philosopher's stone and on the at-
> tendant chemical inquiries were, as a rule, not
> doctors, and carried on their work apart from the
> universities and medical schools. Many, very many
> of them were monks, or at least ecclesiastics, and
> pursued their investigations in solitude and re-
> tirement. A certain antagonism arose between this
> nascent science of chemistry and that older biologi-
> cal learning which formed the basis of medical edu-
> cation. The latter was the heritage of a long-
> established, powerful profession; the former was
> the product of amateurs, of the efforts of scattered
> independent workers, and as such was despised by
> professional men.[45]

Paracelsus' relations with the medical profession were not enhanced by
his assumption of the name "Paracelsus," "greater than Celsus," (the
first century A.D. Roman who is widely considered to have been the
greatest medical mind of antiquity), nor by his actions, on being
named Professor of Medicine and City Physician at Basle, of publicly
burning the works of Avicenna and Galen.[46]

Johannes Baptiste van Helmont (1579-1644), was in many ways Paracelsus' intellectual heir. He was a man of many parts, a respected
scientist who is best remembered for having coined the word, and the
concept, 'gas,' and for having performed a famous experiment with a
willow tree which led him to the quite reasonable, but erroneous, conclusion that all vegetable matter was composed only of water. Van
Helmont was much impressed by the chemical action of fermentation and
attempted to make it the basis of his system of physiology. He was a
leading figure in the school of iatrochemistry. In 1621 he published
a work, *De Magnetica Vulnerum Curatione*, in which he argued for the

curative potency of magnetic forces. This work, with its apparent denial of traditional medicine on the one hand, and the advocacy of a
physical basis for miraculous or spiritual cures on the other, brought
him into conflict with both the physicians and the clergy, and he
spent a good portion of the latter part of his life under house arrest
for his ideas.

Athenasius Kircher (1602-1680), a Jesuit priest, was the first to
describe experiments in hypnotism which were performed on animals.
His research, reported in 1630 in a work titled *Kircher's Experimental
Physiology*, included placing chickens in a somnambulistic state.

Thus, in raising the ghosts of Paracelsus, van Helmont and Kircher, the commissioners were both accurately pointing out some of Mesmer's intellectual forebears and, at the same time, placing him in a
medically and scientifically unsavory tradition. The most influential medical force in eighteenth century Europe was the Dutch physician Hermann Boerhaave (1668-1738), under whose students Mesmer had
studied at Vienna. Boerhaave was an implacable foe of the Paracelsus-
van Helmont tradition. His influence in the medical world is illustrated by the case of William Cullen. About the time Mesmer was publishing his propositions, Cullen, Professor of the Theory of Medicine
at the University of Edinburgh, questioned Boerhaave's opposition to
a chemical basis for medicine. Cullen was ridiculed by his colleagues
as "a Paracelsus, a Van Helmont, a whimsical innovator," and he was
reprimanded for differing from Boerhaave.[47] While Karl Popper may argue that pedigrees have little bearing on truth, they would appear, in
practice, to have a major bearing on the judgment that a particular
system is a pseudoscience.

The medical profession, therefore, had a number of reasons for
dismissing animal magnetism other than those having to do with logical
and empirical demonstrations. First, animal magnetism was a possible
economic threat, it was undermining the hard won monopoly which the
medical profession had established. Second, it was linked to a tradition to which the medical profession had a long standing antipathy
and which it considered to be one of quackery and pseudoscience, to
names still used in the eighteenth century to frighten children
stumbling in the alchemical darkness. (Animal magnetism, of course,
was quackery and pseudoscience within a particular view of the world,
and that fact should not be forgotten. But a great deal of eighteenth
century medicine was quackery and pseudoscience, and could produce a

lot less interesting phenomena than could animal magnetism.) Third,
it was associated with a tradition which stressed the chemical understanding of bodily processes, a point of view generally out of favor
with the medical authorities of the time.

It should be noted in passing that simply because Paracelsus, van
Helmont and Mesmer may have been wrong in their views on healing, that
did not make their opponents right. Just as the detractors of Paracelsus and van Helmont failed to realize the importance of a chemical
understanding of bodily processes, so also the detractors of Mesmer
failed to understand the importance of hypnosis and suggestibility.
This same situation has occurred in other areas of science/pseudo-
science controversy. Where clear distinctions are not or cannot be
made, then not only will legitimate science enhance the reputation of
a closely related pseudoscience, but the pseudoscience can cloud the
reputation, and hide the real value, of the legitimate science. Thus,
the unsavory reputation of phrenology overshadowed, for a time, the
quite legitimate research in the localization of brain function.

What we have been discussing here are examples of the political,
social, economic and professional factors which enter into the judgments of individuals at the time they attempt to make a decision on
the status of a particular system of thought as a pseudoscience. Mesmer seems to have had everything going against him. He was a potential
economic threat and was linked to a tradition which was offensive to
the government, the church and the medical profession. However, a
crucial point with scientists was also that he chose his scientific
metaphors badly, suggesting that his system could be tested in ways
which, in fact, were irrelevant to his claims. This is not an apology
for Mesmer. As I have already stated, I believe his system to have
been a full-blown pseudoscience, internally inconsistent and completely untestable. It is, however, a plea for us to understand what we
are doing. No matter how much we might like to believe that absolute
demarcation criteria exist, the science/pseudoscience distinction is
not a simple, straight-forward one, nor is it a wholly logical one,
nor does it depend exclusively on empirical evidence. It is a judgment influenced by intellectual, social, political, economic, theological and scientific views of a particular time and place. So it is
with all human judgments, science is no exception.

FOOTNOTES

1 For a discussion of the traditional demarcation criteria (justification, validation, falsification, etc.,) see Imre Lakatos, "Falsification and the Methodology of Scientific Research Programmes," in
   Imre Lakatos and Alan Musgrave (Eds.) *Criticism and the Growth of
   Knowledge*, Cambridge: Cambridge University, 1970, pp. 91-196. Possibly the most influential criterion of demarcation for a scientific system has been that of falsifiability, as put forward by Karl
   Popper.

   > Now in my view there is no such thing as induction. Thus
   > inference to theories from singular statements which are
   > 'verified by experience' (whatever that may mean), is
   > logically inadmissible. Theories are, therefore, *never*
   > empirically verifiable. If we wish to avoid the posi-
   > tivist's mistake of eliminating, by our criterion of
   > demarcation, the theoretical systems of natural science,
   > then we must choose a criterion which allows us to admit
   > to the domain of empirical science even statements which
   > cannot be verified.

   > But I shall certainly admit a system as empirical and
   > scientific only if it is capable of being *tested* by
   > experience. These considerations suggest that not the
   > *verifiability* but the *falsifiability* of a system is to
   > be taken as a criterion of demarcation. In other
   > words: I shall not require of a scientific system
   > that it shall be capable of being singled out, once
   > and for all, in a positive sense; but I shall require
   > that its logical form shall be such that it can be
   > singled out, by means of empirical tests, in a nega-
   > tive sense: *it must be possible for an empirical sci-
   > entific system to be refuted by experience.* (*The Logic
   > of Scientific Discovery* New York: Basic Books, 1959,
   > pp. 40-41).

   For Popper, the essential criterion for a scientific theory is that
   it be testable and refutable, at least in principle, through experience. Although his criterion has been widely questioned,
   Popper has reiterated his view on a number of occasions. Thus,

   > One can sum up all this by saying that *the criterion of
   > the scientific status of a theory is its falsifiability,
   > or refutability, or testability.* (*Conjectures and Refu-
   > tations* New York: Harper & Row, 1963, p. 37).

   One consequence of Popper's approach to science is the conclusion
   that it does not matter where the system of ideas to be tested
   comes from, that is, what are the psychological processes or intellectual antecedents which lie behind the formulation of the theory.
   A clear distinction is maintained between the process of discovery
   or formulation and the process of testing or refutation.

   > So my answer to the questions 'How do you know? What is
   > the source or the basis of your assertion? What obser-
   > vations have led you to it?' would be: 'I do *not* know:

> my assertion was merely a guess. Never mind the
> source, or the sources, from which it may spring -
> there are many possible sources, and I may not be
> aware of half of them; and origins or pedigrees
> have in any case little bearing upon truth. But
> if you are interested in the problem which I tried
> to solve by my tentative assertion, you may help
> me by criticizing it as severely as you can; and if
> you can design some experimental test which you think
> might refute my assertion, I shall gladly, and to the
> best of my powers, help you to refute it.' (*Ibid.*, p. 27).

Of course, an historian of science might argue that if actively aiding one's critics to disprove one's most cherished theories is a
necessary criterion for being a true scientist, then the world has
seen few, if any, true scientists. Psychologists might point to a
substantial body of evidence which shows that public adherence to a
particular position leads to increased resistance to change that
position, or accept evidence contradictory to that position as being valid or pertinent.

2 I realize the difficulties associated with the use of "correct" and
   "incorrect" as descriptive terms for science. They will clearly
   suggest to some readers a verificationist position which I do not
   wish to take here. Falsificationist alternatives might be "non-
   falsified" and "falsified." Somewhat more neutral, but not entirely so, might be "tested positively" and "tested negatively." I
   have decided to use "correct" and "incorrect" (retaining the quotation marks) because I believe that these terms are psychologically accurate in describing how many people, even some philosophers,
   actually think about non-rejected and rejected systems of thought.
   In so doing, I am not intending to suggest any acceptance of a
   verificationist position.

3 'Pseudo' is here taken to mean 'false,' 'pretended,' 'spurious' or
   'sham.' 'Proto' will be used in the sense of 'primitive' or 'early.' This is meant to suggest that, in practice, we treat the
   system in question as a primitive science in that we attempt to
   reformulate it or develop it further. It is *not* meant to imply
   that the system in question *must* necessarily be capable of achieving the status of a real science, only that we believe it *may* be
   capable of such achievement. Whether a system of thought is a
   protoscience or a pseudoscience is evidenced by the behavior of
   individuals towards that system of thought at a given time. The
   actual application of the terms 'proto' and 'pseudo' to the system
   of thought by contemporaries is not necessary.

4 Max Black, *Models and Metaphors*, Ithaca, New York: Cornell University, 1962, p. 25.

5 Robert Steinberg, "Component Processes in Analogical Reasoning,"
   Psychological Review, 1977, vol. 84, no. 4, p. 353.

6 Morse Peckham, "Foreword I" to Colin Murray Turbayne's, *The Myth of
   Metaphor*, (Rev. ed.) Columbia, S. C.: University of South Carolina,
   1970, pp. xi-xii.

7 J. Robert Oppenheimer, "Analogy in Science," *American Psychologist*,
   1956, vol. 11, pp. 129-130.

8 Black, pp. 38 ff.

9 *Ibid.*, p. 29.

10 *Ibid.*, p. 37.

11 *Ibid.*, pp. 44-45.

12 *Ibid.*, p. 237.

13 *Ibid.*, pp. 45-46.

14 Mary Hesse, *Models and Analogies in Science*, Notre Dame, Ind.:
    Notre Dame University, 1966, p. 157.

15 *Ibid.*, p. 158.

16 *Ibid.*, p. 160.

17 *Loc. cit.*

18 See, for example, W. H. Leatherdale, *The Role of Analogy, Model and
    Metaphor in Science*, Amsterdam: North-Holland Publishing, 1974.

19 For a discussion of the latter see, Howard E. Gruber, "Darwin's
    'Tree of Nature' and Other Images of Wide Scope," in Judith Wechsler
    (Ed.) *On Aesthetics in Science*, Cambridge, Mass.: M.I.T. Press,
    1978, pp. 121-140.

20 Hesse, pp. 164-165. These comments as well as those of Black raise
    the question of the *psychological* uniqueness of metaphor. A grow-
    ing amount of research on cognitive aspects of metaphoric thought
    exists. On the one hand, the widespread and relatively easy use of
    metaphor in everyday cognitive processes has argued against viewing
    it as a special or deviant mode of thought. On the other hand,
    there do seem to be differences between metaphoric thought and
    other cognitive processes (such as the greater reliance of metaphor-
    ic thought on imagery which has led some psychologists to see meta-
    phor as a kind of bridge between cognition and perception). The
    question, to coin a phrase, requires further investigation.

21 Franz Anton Mesmer, "Dissertations on the Discovery of Animal Mag-
    netism," in Maurice M. Tinterow (Ed.), *Foundations of Hypnosis From
    Mesmer to Freud*, Springfield, Ill.: Charles C. Thomas, 1970, p. 35.

22 *Ibid.*, p. 54.

23 *Ibid.*, p. 55.

24 *Ibid.*, p. 56.

25 *Ibid.*, p. 55.

26 Richard S. Westfall, *Force in Newton's Physics*, London: MacDonald,
    1971, p. 331.

27 Frank Podmore, *From Mesmer to Christian Science*, New Hyde Park,
    N. Y.: University Books, 1963 (orig. pub. in 1909), p. xix.

28 Benjamin Franklin, *et al.* "Report of the Commissioners Charged by
    the King to Examine Animal Magnetism," in Maurice M. Tinterow (Ed.),
    Foundations of Hypnosis From Mesmer to Freud, Springfield, Ill.:
    Charles C. Thomas, 1970, pp. 89-90.

29 *Ibid.*, p. 86.

30 *Ibid.*, p. 114.

31 *Ibid.*, pp. 122-123.

32 *Ibid.*, p. 123.

33 *Ibid.*, p. 124.

34 Mesmer, p. 55.

35 *Loc. cit.*

36 *Loc. cit.*

37 *Loc. cit.*

38 *Loc. cit.*

39 Franklin, p. 125.

40 Alfred Binet and Charles Féré, *Animal Magnetism*, New York: D.
    Appleton, 1896, pp. 17-18.

41 Theodore R. Sarbin, "Attempts to Understand Hypnotic Phenomena," in
    Leo Postman (Ed.), *Psychology in the Making: Histories of Selected
    Research Problems*, New York: Alfred A. Knopf, 1962, pp. 745-746.

42 Quoted in Marc Bloch, *The Royal Touch* (Trans. by J. E. Anderson),
    Montreal: McGill-Queen's University Press, 1973, p. 29.

43 Franklin, p. 123.

44 Sir Michael Foster, *Lectures on the History of Physiology During
    the 16th, 17th & 18th Centuries*, New York: Dover, 1970, (Orig.
    pub. 1901), p. 124.

45 *Ibid.*, pp. 125-126.

46 Avicenna was one of the great Arab physicians of the tenth and
    eleventh centuries. Galen was a second century anatomist and phy-
    sician whose writings made up a portion of the influential Aristo-
    telian corpus of knowledge.

47 Lester S. King, *The Medical World of the Eighteenth Century*, Hunt-
    ington, N. Y.: Robert E. Krieger, 1958, pp. 59-60.

LEGAL SCIENCE AND LEGAL JUSTIFICATION

The demarcation problem in philosophy, even viewed just in the narrow context of recent philosophy of science, has taken many forms. Although Popper was concerned to draw a line between science and non-science - metaphysics, pure logic and mathematics on the one hand, and pseudosciences (as he thought) such as Marxism and psychoanalysis on the other - the verifiability criterion of the logical positivists was generally thought to have a wider scope, being used rather to separate science and scientific philosophy from metaphysics, ethics, aesthetics and other such enterprises.[1] Popper, of course, has stressed that the positivists were anxious to formulate a criterion of cognitive meaningfulness, whereas he was interested only in drawing a line of demarcation; nevertheless, it is not implausible to view the positivists as being in part interested in demarcation as well, in which case science was supposed to be capable of being distinguished in some clear way from other fields which did not have empirical content.

Now, the topic of this Symposium suggests that, insofar as we are concerned with a problem of demarcation, it is the problem of the demarcation between science and pseudoscience. Is there some clear set of criteria by means of which we can place astrology or alchemy clearly in with the pseudosciences and astronomy or chemistry with the sciences? And what shall we say of fields such as psychoanalysis or sociobiology? Which category do they belong in, and by what criteria? Perhaps the only agreement easily achieved on these questions is that they have no simple, ready, generally-subscribed-to answers. But if the issue of where to draw a line of demarcation between science and pseudoscience is a vexed one, the problem of how to distinguish between science and certain other enterprises such as art, humanistic research, or law is no easier to solve.

Traditionally it has been thought that there was a methodology unique to science, and that the use of this method was common to scientific practice and absent from the practice of pseudoscience. Of course, it has been a matter of some controversy just what the scientific method consists in, but common features have usually been taken to be some version of inductive inference, and the testing of hypotheses. Recently it has become fairly standard to view scientific

inference as a form of inference to the best explanation where we select from among the plausible alternative hypotheses the best (according to some criteria which I shall not discuss here), perhaps by eliminating the least adequate among them.[2] Needless to say, I do not
wish to claim that science can be clearly distinguished from pseudoscience on the basis of its use of this method: it may be, for example,
that a distinction will require reference to some essentially social
criteria, or even that no clear distinction can be drawn at all, in
spite of the fact that we have wide agreement on what falls into each
category. Nevertheless, I want to claim that, whatever is true of
pseudoscience, scientific inference *can* be viewed as inference to the
best explanation, properly construed.

But pseudoscience aside, does this provide us with any way of
distinguishing between science and certain social pursuits that are not
typically thought of as scientific? We are, after all, concerned with
science, pseudoscience and *society*, and the latter should not be construed so narrowly as to suggest an interest only in the question of
whether there are essentially social criteria for what is to be identified as science. In this paper I address a different aspect of the
science-society axis - the question of the relation between justification in science and in law. I want to argue that methodological criteria do not provide the touchstone for distinguishing between science
and law, and that, in certain fundamental respects the pattern of justification in these two areas is the same.

This is not, of course, to say that law is a science. Rather, I
want to point to important methodological similarities and to suggest
that, if a line of demarcation is to be drawn, it must be on other
grounds. Nonetheless, it is interesting to observe that law has itself been thought to be a science, and that this idea is not particularly new. Indeed, it has even been argued that law, in the form in
which it emerged in the twelfth century, is the first modern western
science.

> A science in the modern Western sense...focuses on
> formulating hypotheses that can serve as a basis
> for ordering phenomena in the world of time, and
> hence in the world of probabilities and predictions
> rather than certitudes and necessities. The legal
> science of the scholastic jurists was just that
> kind of science. ...It was not, to be sure, an
> "exact" science, like modern physics or chemistry,
> nor was it susceptible to the kind of laboratory
> experimentation that is characteristic of many

> (though not all) natural sciences, although it
> did...utilize its own kinds of experimentation.
> Also it was concerned with constructing a system
> out of observed *social* phenomena - legal institu-
> tions - rather than observed phenomena of the
> world of matter. Nevertheless, like the natural
> sciences that developed in its wake, the new
> legal science combined empirical and theoretical
> methods.[3]

On this view, a science is to be identified in relation to three
sets of criteria - methodological, value, and sociological, the most
important of these for my purposes being the methodological ones.

> A science in the modern Western sense may be de-
> fined in methodological terms as: (a) an inte-
> grated body of knowledge (b) in which particular
> occurrences or phenomena are systematically ex-
> plained (c) in terms of general principles or
> truths ("laws"), (d) knowledge of which (that is,
> of both the phenomena and the general principles)
> has been obtained by a combination of (i) obser-
> vation, (ii) hypothesis, (iii) verification, and
> (iv) to the extent possible, experiment. However,
> (e) the scientific method of investigation and
> systematization, despite these common characteris-
> tics, is not the same for all sciences but must be
> specifically adapted to the particular kinds of
> occurrences or phenomena under investigation by
> each particular science.[4]

The central feature in all this is the notion of explanation and
theory building. For example, late eleventh and twelfth century jurists noted that, in the various legal systems they examined "the question arose whether one who was forcibly dispossessed of his goods had
the right to take them back by force."[5] In Roman law we find that
someone forcibly dispossessed of land may not take it back by force
after a certain time has elapsed. This rule is applied by certain
church councils also to goods, and eventually the general hypothesis
emerges that people whose rights are violated may vindicate them by
resort to legal action but not by "taking the law into their own
hands."[6] So the jurists were engaged in formulating hypotheses which
they then tested against actual decisions and legal rules. It is a
process very like this that some contemporary legal theorists have
fastened on as the paradigm for legal justification, and it is a discussion of this paradigm and its relation to scientific reasoning that
is the main concern of my paper.

Now admittedly, the conception of science involved in this account is rather general, and one might say that, far from being specific to science as we know it, it applies to almost any intellectual

enterprise. This may well be so. But it is part of my point (though I
cannot take time to explore it here) that it is difficult to characterize science more precisely in such a way as to exclude other fields
without at the same time excluding much of what we should want to *in-*
clude. If this is so, then we may have to be content with a characterization of science that does not differentiate it from pseudoscience,
from law, or even from literary criticism,[7] though this is not to say
that any of these fields *is* science. In any case, the issue in contemporary legal theory is not so much whether law is a science as it is
whether legal justification follows a scientific model. I turn, then,
to the theory of the justification of judicial decisions, and specifically to the Rights Theory of Ronald Dworkin.[8] Dworkin's is surely the
most discussed current view, and it is especially interesting for my
purposes because Dworkin makes an explicit comparison between his characterization of judicial justification and justification in two other
areas - in science and in literary criticism. I want to challenge some
of what he says both about justification in each of these areas, and
about the relations among them, thereby, I hope, throwing some light on
the nature of both judicial and scientific reasoning.

Dworkin's theory of justification is, of course, complex, and I
can offer only a very brief and I fear, inadequate summary here. His
main focus is on the question of how we are to come to decisions in
hard cases in civil law - cases, that is, where no clear rule deter-
mines whether plaintiff or defendant should win. He distinguishes in
such cases between arguments of principle and arguments of policy,[9] and
only the former, which justify decisions in terms of individual or
group *rights* (rather than in terms of some collective goal of the community) may be used in the justification of precedent. He also invents
a philosophical judge of "superhuman skill, learning, patience and
acumen"[10] whom he calls Hercules and who constructs a scheme of principles that best justifies the precedents. Hercules

> must discover principles that fit, not only the
> particular precedent to which some litigant di-
> rects his attention, but all other judicial de-
> cisions within his general jurisdiction and, in-
> deed, statutes as well, so far as these must be
> seen to be generated by principle rather than
> policy.[11]

There are two problems that Hercules will face. The first is to
decide what weight to give to the actual arguments the judges used in
deciding the precedent-cases. Here Dworkin is quite forthright. If

a principle other than the one the judge cited both justifies the case
in question and also provides a smoother fit with the arguments taken
to justify other related cases, then this new principle should be
adopted.   Dworkin cites Brandeis and Warren's famous argument about the
right to privacy:

> they argued that this right was not unknown to the
> law but was, on the contrary, demonstrated by a
> wide variety of decisions, in spite of the fact
> that the judges who decided these cases mentioned
> no such right.[12]

The theory of justification is thus directed not merely to the descriptive question of what justification was actually offered for a particular decision: it also addresses the normative question of what would
be the best justification, in some sense of 'best' yet to be explained.

The second problem faced by Hercules is that the requirement of
consistency with all previous decisions is clearly too strong. No set
of principles can reconcile all standing statutes and precedents.
Some part of 'settled' law will have to be re-interpreted as mistaken,
but this must be a principled re-interpretation, else the consistency
requirement has no real force. The details of how this is done need
not concern us: it is the general technique that is important. What
Hercules must do is to engage in what might be described (by analogy
with inference to the best explanation) as inference to the best theory, in the sense that his justification is stronger than any plausible
alternative that fails to recognize any mistakes, or that recognizes a
different set of mistakes. He may, for example, find on balance that
he must reach a decision that opposes popular morality on some issue -
say, abortion. Here he judges that the community's morality is inconsistent, and he must opt for the decision that coheres with whatever
general theory does the least violence to the whole system including
some reasonably coherent notion of what morality requires. What this
amounts to in particular cases is, of course, highly controversial.
But Dworkin insists that the fact that well-informed and responsible
judges disagree about what the right answer is goes no way at all toward showing that there is no right answer. Indeed, he holds that in
every civil case there is some particular truth of the matter, and one
of the parties has a *right* to win, even though the decision may be
controversial, and seriously underdetermined by the black letter rules.

So we have the outlines of a coherence theory of judicial justification. Of course no merely human judge could be expected to make

decisions in this way, but Dworkin clearly believes his characterization can serve at least as a model to which judges can and ought to
achieve some more-or-less close degree of approximation.

Theories of justification that are, broadly speaking, coherence
theories, are, of course, common in many areas of philosophy. John
Rawls' notion of reflective equilibrium as a method for developing a
theory of justice provides one such theory:

> By going back and forth, sometimes altering the
> conditions of the contractual circumstances, at
> others withdrawing our judgments and conforming
> them to principle, I assume that eventually we
> shall find a description of the initial situa-
> tion that both expresses reasonable conditions
> and yields principles which match our consider-
> ed judgments duly pruned and adjusted.[13]

Now, Rawls tells us further that the "process of mutual adjustment of principles and considered judgments is not peculiar to moral
philosophy."[14] In this connection he refers us to Nelson Goodman's
discussion of the justification of principles of inductive and deductive inference. The classic formulation of the sort of principle Rawls
has in mind occurs in this passage from *Fact, Fiction and Forecast*:

> The point is that rules and particular inferences
> alike are justified by being brought into agree-
> ment with each other. A rule is amended if it
> yields an inference we are unwilling to accept:
> *an inference is rejected if it violates a rule we
> are unwilling to amend*. The process of justifi-
> cation is the delicate one of making mutual ad-
> justments between rules and accepted inferences....[15]

Joel Feinberg has compared this general sort of characterization
of philosophical method with the kind of reasoning that goes on in
courts of law:

> General principles arise in the course of delibera-
> tions over particular problems, especially in the
> efforts to defend one's judgments by showing that
> they are consistent with what has gone before. If
> a principle commits one to an antecedently unac-
> ceptable judgment, then one has to modify or sup-
> plement the principle in a way that does the least
> damage to the harmony of one's particular and gen-
> eral opinions taken as a group. On the other hand,
> when a solid, well-entrenched principle entails a
> change in a particular judgment, the overriding
> claims of consistency may require that the judgment
> be adjusted. This sort of dialectic is similar to
> the reasonings that are prevalent in the law courts.
> When similar cases are decided in opposite ways, it
> is incumbent on the court to distinguish them in
> some respect that will reconcile the separate

> decisions with each other and with the common
> rule applied to each. Every effort is made to
> render current decisions consistent with past
> ones unless the precedents seem so disruptive
> of the overall harmony of the law that they
> must, reluctantly, be revised or abandoned.[16]

Now, according to Rawls, moral philosophy and "the study of other
philosophical principles such as those of induction and scientific
method"[17] provide but a few examples of a widespread phenomenon. The
same sort of method is evident in linguistics when we try to describe
the "sense of grammaticalness that we have for the sentences of our
native language"[18]: our linguistic theory can be expected to affect to
some degree that sense of grammaticalness, even if it can hardly be expected to effect "a substantial revision of our sense of correct gram-
mar."[19]

So there is only a difference of degree between the sort of method
appropriate in moral philosophy and what is suitable for theoretical
linguistics. But when it comes to the natural sciences we find a
rather sharp contrast. Thus, to take what Rawls refers to as an extreme case,

> ...if we have an accurate account of the motions
> of the heavenly bodies that we do not find appeal-
> ing, we cannot alter these motions to conform to a
> more attractive theory. It is simply good fortune
> that the principles of celestial mechanics have
> their intellectual beauty.[20]

The suggestion seems to be that in the moral sphere we *can* alter whatever the analogues of the motions of the heavenly bodies are, in the
interests of a better theory. And this implies that Rawls believes
there is an important difference between the sort of methodology appropriate in philosophical theorizing, even when that theorizing is about
science, and scientific theorizing itself.

The view that there is this difference is espoused also by Dworkin
in his discussion of Rawls' method in "Justice and Rights."[21] Dworkin
describes two general models - the natural or scientific model and the
constructive model - that define the kind of coherence meant to be
achieved by the technique of reflective equilibrium, and gives reasons
for saying that the Rawlsian account requires the constructive model
and, indeed, is incompatible with the natural one. Pretty clearly he
means these observations to apply to his own "Rights Thesis" as well,
and thus to be central in the correct account of the justification of
judicial decision. Yet some of the ways in which Dworkin elaborates

his thesis about legal methodology suggest a close analogy with science
itself. Thus he says

> I oppose the popular theory that judges have dis-
> cretion to decide hard cases. I concede that
> principles of law are sometimes so well balanced
> that those favoring a decision for the plaintiff
> will seem stronger, taken together, to some law-
> yers but weaker to others. I argue that even so
> it makes perfect sense for each party to claim
> that it is entitled to win, and therefore each
> to deny that the judge has a discretion to find
> for the other....I do not claim (indeed I deny)
> that the process of decision will always yield
> the same decision in the hands of different
> judges. Nevertheless I insist that *the process,
> even in hard cases, can sensibly be said to be
> aimed at discovering, rather than inventing, the
> rights of the parties concerned....*[22]

There is something noble at stake in this conception, according to
Dworkin, and it has to do with our overall view of the political process. He says:

> Citizens are encouraged to suppose that each has
> rights and duties against other citizens, and
> against their common government, even though these
> rights and duties are not all set out in black-
> letter codes. They are therefore *encouraged to
> frame and test hypotheses about what these rights
> are*, and to treat one another, and demand to be
> treated by the state, under the beneficial and
> unifying assumption that justice is always rele-
> vant to their claims even when it is unclear what
> justice requires.[23]

This basic theme that rights are to be discovered rather than invented, and that this is to be done by some form of hypothesis testing
is both characteristic of Dworkin's recent work and analogous to a very
common conception of scientific methodology. Dworkin tells us that we
have heretofore had a mistaken notion of what law is, for we have
thought of the expression 'existing law' as referring to some sort of
independent entity. In contrast, he hopes, he says

> to persuade lawyers to lay the entire picture of
> existing law aside in favor of a theory of law
> that takes questions about legal rights as special
> questions about political rights, so that one may
> think a plaintiff has a certain legal right with-
> out supposing that any rule or principle that al-
> ready 'exists' provides that right. In place of
> the misleading question, whether judges find rules
> in the 'existing law' or make up rules not to be
> found there, we must ask whether judges try to en-
> force the rights they think parties have, or whether
> they create what they take to be new rights to serve
> social goals.[24]

Now, from the point of view of the ontology, as it were, of the situation, one is at first led to wonder what difference there can be between believing that there is existing law and believing that there are rights which can be discovered. The picture sounds very like the one implicit in Dworkin's idea of the natural model of reflective equilibrium wherein theories of justice (and presumably theories of rights?) "describe an objective moral reality; they are not, that is, created by men or societies but are rather discovered by them, as they discover laws of physics."[25] But it soon becomes clear that the discovering of principles in the law, as opposed to creating them, is not meant by Dworkin in the scientific sense. Legal principles do not mirror some independent reality in the way that scientific laws do, he might say, for whereas scientific laws are descriptive, legal ones are fundamentally prescriptive. Legal reasoning, instead, can be captured in the other - the constructive - model. This second model, he says,

> treats intuitions of justice not as clues to the
> existence of independent principles, but rather
> as stipulated features of a general theory to be
> constructed, as if a sculptor set himself to
> carve the animal that best fits a pile of bones
> he happened to find together. This 'constructive'
> model does not assume, as the natural model does,
> that principles of justice have some fixed, ob-
> jective existence, so that descriptions of these
> principles must be true or false in some standard
> way. It does not assume that the animal it
> matches to the bones actually exists.[26]

It is instructive that Dworkin pictures a sculptor here rather than, say, an archaeologist. An archaeologist might make the same carving as the sculptor, but perhaps with something else in mind. What becomes clear as one thinks about the analogy is the centrality of the notion of 'best fit.' A sculptor might be satisfied with just an interesting or beautiful animal-like shape, whereas the archaeologist will insist on an outcome at least consistent with what else we know about animals that existed at the relevant time and place. She will, in short, construct a figure that provides the best fit not only with these bones, but with other things we know: she may even decide that some of the bones that belong with these are missing, or, on the other hand, that some we have here belong elsewhere. But what the archaeologist does fits the natural model on Dworkin's view, rather than the constructive. On the natural model "[m]oral reasoning or philosophy is a process of reconstructing the fundamental principles by assembling

concrete judgments in the right order, as a natural historian reconstructs the shape of the whole animal from the fragments of its bones
that he has found."[27]

On the constructive model, Dworkin seems to be saying, we seek
'best fit' in the sense of consistency. But the theory thus constructed does not have the additional constraint he takes to be imposed on
theories in science - of a tie somewhere to objective reality, either
in the form of observation sentences, or predictions or some such
thing. So there is an important difference between the two models,
for Dworkin, which can be summed up in the remark that on the scientific model observation, or its moral analogue, intuition, is primary,
whereas on the constructive model the theory is primary. The moral
analogues of observations, moral intuitions, are, on Dworkin's conception of the natural model, produced by a moral faculty that is the instrument for discovering an objective moral reality. "These intuitions
are clues to the nature and existence of more abstract and fundamental
moral principles, as physical observations are clues to the existence
and nature of fundamental physical laws."[28] Suppose a judge finds that
he has some intuition - one of Dworkin's examples is the intuition that
a particular minority is entitled to special protection - that he cannot by any consistent set of principles reconcile with his other intuitions, for example, the view that distinctions based on race are in-
herently unfair to individuals. This latter view apparently is intended by Dworkin to represent the judge's considered moral theory. Faced
with this dilemma, Dworkin thinks the scientifically-minded judge will
follow "the troublesome intuition,...submerging the apparent contradiction, in the faith that a more sophisticated set of principles, which
reconciles that intuition does in fact exist though it has not been
discovered."[29] This analysis emerges from an analogy between moral intuitions and observational data and the latter, Dworkin supposes, must
always hold sway in science.

The constructive model, on the other hand, requires that the theory be regarded as primary in the sense that intuitions that just
won't fit the theory must be compromised. The coherence requirement
is an independent requirement of political morality rather than a requirement that flows from the assumption that moral convictions are accurate observations of some actual state of affairs.

There is thus, for Dworkin, a sharp line to be drawn between his
two models. He would grant, I think, that in science there might be

different explanations of the same data, that our theories are underdetermined by the evidence and so there might be more than one way of
'saving the phenomena.' Indeed, he refers here to the views of philosophers like Quine, though he pretty clearly has reservations about
the correctness of such accounts of science. Such philosophers, he
tells us

> suppose that our concepts and our theories face
> our experience as a whole, so that we might re-
> act to recalcitrant or surprising experience by
> making different revisions at different places
> in our theoretical structure if we wish. Re-
> gardless of whether this is an accurate picture
> of scientific reasoning, it is not a picture of
> the procedure of equilibrium, because this pro-
> cedure argues not simply that alternative struc-
> tures of principle are available to explain the
> same phenomena, but that some of the phenomena,
> in the form of moral convictions, may simply be
> ignored the better to serve some particular
> theory.[30]

This is too starkly put, I suspect, from two standpoints. In the
first place, I don't think we really *do* simply *disregard* evidence that
won't fit our moral or legal theory. Dworkin's example here is that
we might be utilitarians and yet think that slavery is unjust in spite
of its leading to greater utility. If even on reconsideration we
still hold both views and cannot reconcile them, the constructive model
authorizes us to ignore the slavery intuition in order to maintain
equilibrium. I think what really happens in a case like this is that
we continue the effort at reconciliation, seeking either to modify the
theory to accommodate the intuition or modify the intuition, perhaps
by finding reasons that it shouldn't be given such great weight. It is
at least unlikely that we'll simply grant that slavery is, after all,
acceptable; instead, we will surely make some other adjustment. For
example, we might try to show that slavery is really different from
what we had supposed, or perhaps even, in an extreme case, we might
continue to maintain incompatible beliefs, recognizing them to be in
conflict, having no means for resolving them, but believing that some
resolution will present itself eventually. Further, the example is
particularly unfortunate in being inapt for Dworkin's purposes, for
utilitarianism does claim to accommodate the slavery intuition by
showing that slavery has long-term bad consequences, so that this is
not even a case where we're seriously tempted to think we may have to
relinquish the uncomfortable intuition.

In the second place, if historians of science are to be believed,
then it's not true that the natural model is so insistent on the pri-
macy of observation. Sometimes we say the test tubes were dirty, or we
make adjustment elsewhere in the system without giving up the theory,
and sometimes we even do what Dworkin says only the constructive model
allows: we say the theory *must* be right and either something's gone
wrong with the observation or we'll soon find a way to accommodate it,
though that way eludes us for the moment. It's true that we don't *sim-*
ply ignore the recalcitrant observation in science, but neither do we
simply forget about the discordant moral intuition or legal decision:
we regard ourselves as bound to explain the discrepancy.

A plausible analogy is the inductive process of curve-fitting.
If there is a point in the data that falls some distance from the best
curve, however measured, there are surely circumstances in which we may
be entitled to disregard it. Furthermore, the curve that best fits
settled law may have to change as more law becomes settled. Theories
that would best justify decisions before and after *Brown v. Board of*
Education[31] are presumably different in some important respects. We
might even go so far as to say there are occasional 'conceptual revolutions' (in Kuhn's sense) in law, as with the introduction of accident
compensation schemes not based on fault.[32] As a matter of fact, Kuhn
is, I suppose, our most obvious source of the view that we do not reject theories in science on the basis of falsifying instances. And
yet, one consults Kuhn in vain for specific examples of situations in
which scientists have straightforwardly ignored or even rejected observational data in favor of a well-entrenched theory; what he gives
us are examples in which anomalies and even crises in the prevailing
scientific theory failed to occasion its abandonment until a plausible
alternative was available. But the offending observations were rarely
just ignored.[33]

There are, however, some interesting examples of what appears to
be just glossing over recalcitrant observations in the history of science. One occurs in Turbayne's account of how Isaac Barrow dealt with
a problem he encountered in optics, involving the question of where an
object seen through converging lenses and mirrors should appear. Barrow argued from the accepted geometrical theory of optics that the object should appear "Extremely remote because diverging rays mean near,
less diverging rays mean less near, parallel rays mean far, therefore,
converging rays mean very far."[34] But when he looked, he found "that

the facts were directly opposite to what they ought to have been."
Turbayne comments:

> Barrow's observation report was good. Although
> he was completely nonplussed by this turn of
> events, he remained undismayed. Having found
> the facts "repugnant" to the theory which, he
> said, "I know to be manifestly agreeable to
> reason", he refused to renounce the theory.[35]

He was sure there was something that needed explaining, but admitted that he was unable to solve the problem. Nonetheless, the existence of the recalcitrant observation apparently didn't move him in
the least to abandon the theory, nor did he even suppose that the ultimate solution to the problem would involve a modification of the theory.[36]

An even more interesting example of this sort of phenomenon is
discussed at some length by Michael Polanyi. He writes,

> The Michelson-Morley experiment of 1887, which
> Einstein mentions in support of his theory and
> which the textbooks have since falsely enshrined
> as the crucial evidence which compelled him to
> formulate it, actually did not give the result
> required by relativity! It admittedly substan-
> tiated its authors' claim that the relative mo-
> tion of the earth and the 'ether' did not exceed
> a quarter of the earth's orbital velocity. But
> the actually observed effect was not negligible;
> or has, at any rate not been proved negligible
> up to this day. The presence of a positive ef-
> fect in the observations of Michelson and Morley
> was pointed out first by W. M. Hicks in 1902 and
> was later evaluated by D. C. Miller as correspond-
> ing to an 'ether-drift' of eight to nine kilo-
> meters per second. Moreover, an effect of the
> same magnitude was reproduced by D. C. Miller
> and his collaborators in a long series of experi-
> ments extending from 1902 to 1926, in which they
> repeated the Michelson-Morley experiment with
> new, more accurate apparatus, many thousands of
> times.

> The layman, taught to revere scientists for
> their absolute respect for the observed facts,
> and for the judiciously detached and purely pro-
> visional manner in which they hold scientific
> theories (always ready to abandon a theory at
> the sight of any contradictory evidence), might
> well have thought that, at Miller's announcement
> of this overwhelming evidence of a 'positive
> effect' in his presidential address to the Amer-
> ican Physical Society on December 29th, 1925,
> his audience would have instantly abandoned the
> theory of relativity. Or, at the very least,
> that scientists - wont to look down from the

> pinnacle of their intellectual humility upon
> the rest of dogmatic mankind - might suspend
> judgment in this matter until Miller's results
> could be accounted for without impairing the
> theory of relativity. But no: by that time
> they had so well closed their minds to any
> suggestion which threatened the new ration-
> ality achieved by Einstein's world-picture,
> that it was almost impossible for them to
> think again in different terms. Little at-
> tention was paid to the experiments, the evi-
> dence being set aside in the hope that it would
> one day turn out to be wrong.[37]

So, at least according to this account, it would appear that there are
cases in science in which the data are disregarded, though this is
probably rare. But it is equally rare in the moral-legal sphere, as I
have argued.

Furthermore, in science we have a related phenomenon which, on
Dworkin's account one would have thought should not appear there. This
is the phenomenon that scientific theories are sometimes replaced by
successor theories, generally thought to be better than their predecessors; and yet the successor theories cannot explain certain facts
that were explained by their predecessors. For example:

> (a) By the 1670's, the celestial mechanics of
> Descartes was widely accepted (in spite, inci-
> dentally, of its failure to offer any explana-
> tion for the precession of the equinoxes - which
> had been a solved problem since antiquity). One
> of the core problems for Descartes, as for Kepler
> before him, was that of explaining why all the
> planets move in the same direction around the
> sun. Descartes theorized that the planets were
> carried by a revolving vortex which extended from
> the sun to the periphery of the solar system.
> The motions of this vortex would entail that all
> the planets moved in the same direction. Newton,
> on the other hand, proposed no machinery whatever
> for explaining the uniform direction of revolu-
> tion. It was perfectly compatible with Newton's
> laws for the planets to move in quite different
> directions. It was acknowledged by both critics
> and defenders of the newer Newtonian system that
> it failed to solve this problem which had been
> explained by the earlier Cartesian system.
> (b) Prior to Hutton and Lyell, geological theo-
> ries had been concerned with a wide range of
> empirical problems, among them: how aqueous
> deposits are consolidated into rocks, how the
> earth originated from celestial matter and
> slowly acquired its present form, when and
> where the various plants and animals originated,
> how the earth retains its heat, the subterraneous

> origins of volcanos and hot springs, how and when
> various mineral veins were formed, etc. Solutions,
> of varying degrees of adequacy, had been offered to
> each of these problems in the 18th century. The
> system of Lyell, and similar ones which largely
> displaced these earlier geological theories by the
> mid-19th century, did not offer *any* explanation for
> *any* of the problems cited above.[38]

This sort of example is not easy to fit into Dworkin's conception of
the natural model, though analogous occurrences in the moral sphere
could presumably be accounted for well enough in the constructive model. But then it looks as though something that occurs with at least
some frequency in science does not conform to what Dworkin conceives to
be the scientific model of theory justification.

Another difficulty with the natural model is said by Dworkin to be
this. The equilibrium technique is relativistic in two ways: in that
(1) it selects the best theory from a small finite list; and (2) the
results are relative to an area of initial agreement. The first species of relativism is said to be a problem because if we think only
that we have the best theory from a small list, and that further work
will yield a better one, then we have very little reason to suppose
that the principles we have arrived at at a given stage are an accurate
description of moral reality. Now I think this begins to make clear
what has gone wrong with the models: it is the supposition that the
natural model requires that we believe our observations put us in touch
with transcendent reality and so the moral version must imply that moral intuitions put us in touch with a transcendent moral reality. But
all of this is surely questionable for, even if we think of science as
ultimately directed toward discovering the truth, it surely does not
follow that we have attained that truth at any particular stage, or
that there is a clear and unassailable correspondence between our observations and certain features of an external reality, as Dworkin apparently believes.

Indeed, it seems patent that Dworkin accepts some fairly naive
version of scientific realism. He refers to an

> astronomer who has clear observational data that
> he is as yet unable to reconcile in any coherent
> account, for example, of the origin of the solar
> system. He continues to accept and employ his
> observational data, placing his faith in the idea
> that some reconciling explanation does exist
> though it has not been, and for all he knows may
> never be, discovered by man.[39]

What this means when applied to the moral or legal sphere is that we're stuck with the data - for example, the intuition that slavery is wrong. Considered moral judgments - intuitions or convictions - are analogous to direct perception reports which are in some important way privileged and immune from revision. But very little of science is built on this sort of foundation, and so from this point of view the constructive model is really a better model for science than is the natural one. However, even this way of putting it gives too much to Dworkin's characterization, for I want rather to claim that the contrast he draws between the two models is too sharp. Insofar as the natural model fits scientific theorizing, it equally well fits the Herculean judicial enterprise; and insofar as the constructive model fits the latter, it equally well fits science. Perhaps it would be more accurate to say that a better picture of the scientific enterprise is a kind of composite of the natural and constructive models. If there are existence assumptions in science they are much weaker than the ones Dworkin attributes to the natural model; and the adjustment between theory and intuition can come out either way, though most often we cling to our theory if it's been carefully thought through and tested - at least until the anomalies pile up, or some better theory comes into view.

As a matter of fact, the first species of relativism that Dworkin attributes to the equilibrium technique - that of selecting the best theory from a small finite list - is characteristic also of scientific inference conceived in a certain way. This conception trades on a fact that has gradually become clear over the period since the heyday of positivist philosophy of science - that we do not have any formal way of providing necessary and sufficient conditions for theory acceptance in science. High probabilities are not sufficient as the lottery paradox shows us,[40] and probably not necessary either, and no other formal or quantitative criterion seems to do the job any better. Instead, it is plausible, as I suggested earlier, to regard scientific inference as inference to the best explanation or theory - a form of eliminative inference, where the eliminations are made from a small finite list of hypotheses or theories.[41] Our theories are always open to revision, and the most we can say of those we now accept is that they are the best we can do for the moment. As to the second form of relativism - that results are relative to an area of initial agreement - it is widely accepted that, in science *and* philosophy, justification *always*

rests on initial commitments.[42] One must start somewhere, though this
does not mean the initial commitments are immune from criticism or revision, or that objectivity cannot be achieved.

Thus I think that most of what Dworkin wants to say about the nature of judicial reasoning is at least consistent with a scientific
model, properly construed. Legal propositions such as 'plaintiff is
entitled to damages' are either true or false, he would insist, though
we may not know which. He says,

> A proposition of law may be asserted as true if
> it is more consistent with the theory of law
> that best justifies settled law than the contrary
> proposition of law. It may be denied as false if
> it is less consistent with that theory of law than
> the contrary.[43]

But we might still want to insist, both in law and in science, that to
be justified in asserting a proposition as true is not to have shown
that it *is* true. Furthermore, it is not clear, despite Dworkin's insistence on the point, that there is a *single* theory that best justifies settled law. Some accepted decisions will have to be ignored,
and there may well be different ways to do this that yield equally acceptable theories from the coherence point of view. Similarly with the
scientific case: here too there may at a particular time be equally
acceptable explanations of what we take to be the facts.

One reason, of course, that we are inclined to say this is that
we are always operating with some measure of ignorance and uncertainty,
and equally rational, sincere, well-trained and intelligent lawyers, or
scientists, often disagree. This fact, and the fact that in the law we
do not and cannot have rules to cover explicitly every conceivable
case, has led many jurisprudents to say that, in hard cases where the
rules run out, judges have discretion to decide as they see fit: there
is no single right answer.[44] If what we mean by a right answer in the
law is one that expresses a true proposition of law - perhaps something
like 'Jones is liable for Smith's injuries' - then, according to Dworkin, the underlying thesis of the 'no single right answer' view must
be that if a proposition is true it is demonstrable as true, presumably on the basis of the facts at our disposal.[45] This is, as one
might expect, a thesis that Dworkin wishes to deny, and to do this he
asks us to consider yet another analogy. We are to imagine a group of
Dickens scholars engaged in a kind of literary game in which they discuss David Copperfield as though he were a real person. They might

play this game with different sets of ground rules which will, naturally, lead to different results.

Consider, for example, a rather ascetic set of ground rules:

> 1) Any proposition about David may be asserted
> as 'true' if Dickens said it, or said something
> else such that it would have been inconsistent
> had Dickens denied it.
> 2) Any proposition may be denied as 'false' if
> Dickens denied it, or said something else such
> that it would have been inconsistent had Dickens
> said it.[46]

Now with these ground rules, there will be many propositions
about David that are neither assertable nor deniable. For example, as
Dworkin tells us

> Dickens never said that David had a homosexual
> affair with Steerforth, and it would not have
> been inconsistent with anything he said if he
> had asserted it. So the participants can nei-
> ther assert nor deny the proposition, not be-
> cause they lack sufficient information, but
> because they have sufficient information to be
> certain that, under their rules, the proposi-
> tion is neither true nor false.[47]

But clearly there can be far less ascetic ground rules, some of
which, Dworkin thinks, define a game very like the actual practice of
literary criticism. In this game,

> a further proposition about David is assert-
> able as true (or deniable as false) if that
> further proposition provides a better (or
> worse) fit than its negation with proposi-
> tions already established, because it explains
> in a more satisfactory way why David was what
> he was, or said what he said, or did what he
> did, according to those already established
> propositions.[48]

In this game it is very likely that all or almost all questions about
David will have a right answer. Those that won't will be trivial, such
as the question whether David had freckles on his toes. And of course
it is clear that Dworkin similarly imagines that, although there can be
different forms of the *legal* enterprise, with more or less strict
ground rules, the form we should have and do have to some extent is the
form that corresponds with the "best fit" version of the literary exercise.

Now, what is interesting, I think, is Dworkin's willingness to buy
this analogy when he was unwilling to accept the scientific one. This
is because, although he thinks both the literary and the legal enterprise depend upon certain facts, these are not the 'hard facts' upon

which the scientific enterprise depends. Instead they are, in the
literary case, what might be called aesthetic facts, such as facts of
narrative consistency which are, however, not even in principle demonstrable by ordinary scientific methods. But surely these facts of
narrative consistency are to at least some extent parasitic on facts
and generalizations about the behavior of ordinary people, which Dworkin acknowledges to be 'hard' facts. For example, we will decide what
behavior David might have engaged in partly on the basis of how people
we know, similar to David in some respects, behave in like actual situations, which is to say, partly on the basis of psychological facts.
Thus it is not so clear that in the literary case we lack any tie to
the external world, even though this tie may be extremely tenuous in
some literary works.

But further, the problem of reconciling the judicial precedents
is much more complex than the David Copperfield game, not only because
we must consider myriads of cases decided by different judges over
long periods of time, but even often in situations where, on the face
of it, the issues appear to be fairly circumscribed. Consider, as an
example, a torts problem that has recently been before the Supreme
Court of Canada, concerning the standard of care required of police
officers in apprehending a suspected criminal, and in particular the
extent to which police officers are justified in using force in effecting an arrest. If the courts are prepared to give clear direction in
these cases this will presumably act as an incentive to police officers
to behave in particular ways, so the potential social importance of
these cases is considerable.

The two cases to be considered are *Priestman v. Colangelo, Shynall
and Smythson*, and *Beim v. Goyer*.[49] The 'facts' are succinctly summarized by Paul Weiler as follows:[50]

> There was a marked similarity in the events....
> Smythson, a 17-year-old in Toronto, and Beim, a
> 14-year-old in Montreal, each stole new cars,
> apparently to take "joy rides". Each was spotted
> by police officers in their vehicles, Smythson by
> Priestman and Beim by Goyer. When the boys re-
> fused to stop, the police gave chase at dangerous
> speeds along the city streets. There the situa-
> tions diverge. Priestman aimed his gun at the
> tire of Smythson's car as he was reaching a busy
> intersection, in order to stop him. Just as he
> fired, his arm was bumped and the bullet acci-
> dentally struck Smythson in the neck, rendering
> him unconscious. The car went out of control,
> ran up on the sidewalk and killed Columbo

Colangelo and Josephine Shynall who were talking
quietly while waiting for a bus. By contrast,
Beim, the driver of the stolen car in the other
case, collided with a parked vehicle and thus
stopped himself. He hopped out of his own car
and fled across a deserted, snowy and rocky
field. Goyer gave chase with his gun in his
hand. He fired two warning shots in the air
and he also fell twice while crossing the field.
Unfortunately, when he tripped and fell a third
time, the gun went off and the bullet hit Beim
in the neck, paralyzing him.

Consider the four possible pairs of outcomes in these cases.

(1) Both policemen are negligent.

(2) Neither policeman is negligent.

(3) Priestman is negligent; Goyer is not.

(4) Goyer is negligent; Priestman is not.

Now on the face of it, one might suppose that the important features
of the cases are sufficiently similar as to dictate either (1) or (2)
as the result. But there are differences. "Priestman *intentionally*
fired his gun in a crowded city area and as a result injured two completely innocent bystanders. Goyer's gun went off *accidentally* while
he was running across a deserted field and as a result injured the es-
caping offender."[51] This points to (3) as the proper result. On the
other hand Beim was not posing any additional threat of injury to members of the community when the mishap occurred, whereas Smythson was
driving recklessly, so one might say it was less important that Beim
be stopped; and Goyer had already fallen during the foot chase, so it
was perhaps negligence for him to go on running with gun in hand.
These features suggest that the result should be (4), as it was in
fact. But the question is whether any *principled* rationale can be
given for this outcome - a rationale that makes it clear why (4) should
be preferred to the alternatives.

Both opinions make reference to section 25(4) of the Canadian
Criminal Code which reads:

A peace officer who is proceeding lawfully to
arrest...any person for an offence...is justified, if the person to be arrested takes flight
to avoid arrest, in using as much force as is
necessary to prevent the escape by flight, unless the escape can be prevented by reasonable
means in a less violent manner.

Now, however much this provision may relieve both Priestman and Goyer
of criminal liability, it certainly does not settle the matter of civil

liability. On that point Mr. Justice Locke argued in the *Priestman*
case as follows:

> The performance of the duty imposed upon police
> officers to arrest offenders who have committed
> a crime and are fleeing to avoid arrest may, at
> times and of necessity, involve risk of injury
> to other members of the community. Such risk,
> in the absence of a negligent or unreasonable
> exercise of such duty, is imposed by the
> statute.... Police officers in this country
> are furnished with firearms and these may, in my
> my opinion, be used when, in the circumstances
> of the particular case, it is reasonably neces-
> sary to do so to prevent the escape of a crimi-
> nal whose actions, as in the present case, con-
> stitute a menace to other members of the public.[52]

The conclusion, of course, was that Priestman was not liable. Neither
the argument that shooting should be considered a last resort, nor
that in this case more harm than good was likely to result from trying
to stop Smythson, especially given the relative lack of seriousness of
his crime, moved the court. But these matters *were* relied upon in a
vigorous and convincing dissent.

Nevertheless, given the *Priestman* decision, it seems clear that
either (a) Goyer should also have been exonerated from liability, or
(b) the Court should have overruled *Priestman* in reaching the contrary
result in *Beim*. Instead, the Court in the latter case held that there
was evidence on which the jury could reasonably base its finding that
Goyer "was at fault in failing to exercise proper care in the use of
fire-arms when pursuing the appellant."[53] In addition, it said:
"Each of the decided cases dealing with the use of fire-arms by peace
officers...turns largely on its own facts."[54] Thus the Court suggested
that the traditional ploy of distinguishing cases on the facts was appropriate in this situation where it desired to uphold the finding
that Goyer was liable, even though Priestman, in what appeared to be
similar circumstances, was not liable.

But there is a difficulty, for the decision in *Beim* seems to be a
reversal of that in *Priestman*, and yet both decisions stand. It is
hard to see how this state of affairs can be thought to provide any
reasonable guidance in similar situations either to the police, to injured citizens, or to the courts, for although distinguishing cases on
the facts is a common judicial technique, it is just not clear that
the factual differences in these two cases are *relevant* factual differences sufficient to justify different outcomes. At the very least,
we need to be given more reason to think that the facts in the two

cases are relevantly different; and without such reason one could
hardly be blamed for thinking the differences to be unimportant from
the standpoint of attempting to construct a coherent legal theory. If
this is so, then on Dworkin's theory and many other theories of ad-
judication, we have a situation in which at least one of the precedents should be viewed as mistaken, and the one that stands should be
the one that coheres best with the rest of settled law. Which result
actually does this I am not prepared to say, though I think strong
reasons of principle and policy can be given for holding the policemen
liable in both cases. No doubt this way of putting the matter would
be too explicitly result-oriented for Dworkin; and in addition the
stumbling block of the Canadian Supreme Court's reluctance to overrule
itself shouldn't be underestimated. But at least for it to have done
so on this matter seems compatible both with our ordinary sense of
fairness about this kind of case and with Dworkin's theory, so far as
its abstractions can be made to apply to such a concrete situation.
And if the theory does apply to this kind of case and, as well, to the
David Copperfield case, it is no less a propos, I submit, for a typical example of small scale scientific theorizing.

        Another problem with the law-science separation is that Dworkin's
use of 'truth' and 'fact' in both the literary and the legal realm is
somewhat peculiar, for these terms used in these contexts do not have
the meanings he apparently supposes them to have in science. Truth in
law and literature is wholly a matter of coherence, on his view, whereas truth in science also involves correspondence with reality. It is
difficult, then, not to suspect Dworkin of embracing paradox for its
own sake, and it is equally difficult not to wonder what is the point
of the talk about discovery and truth and right answers in the law,
since on his account it would seem to bear little relation to such
talk elsewhere. My own view, which I do not have space to develop
here, is that the notions of discovery and fact and even truth in science are more metaphorical than Dworkin's picture allows,[55] so the
sense of paradox his use of these concepts generates is more apparent
than real. This, of course, is just an extension of my earlier argument that Dworkin's contrast between the natural and constructive models of equilibrium is overdrawn.

        But I shouldn't want to convey the impression that I think there
is nothing to Dworkin's insistence on a separation between the legal
and scientific models of reasoning. In a recent Massachusetts case[56]

the Supreme Judicial Court decided that, in a will which directed the
establishment of a scholarship fund "to aid and assist worthy and am-
bitious young men to acquire a legal education," the phrase "young men"
must be construed in its generic sense which includes women. In justification of this construction, the court cited the following facts,
among others: (1) the fund was to be the Richard W. and Florence B.
Irwin Scholarship Fund; (2) it was suggested that the trustee use the
educational program of the Knights Templar as a model, and this program provides financial assistance to students without regard to their
sex; (3) the testator wanted to help "as many students as possible,"
and limiting the applicant pool is inconsistent with this objective.
The court concluded that the testator had no wish to confine his bounty
to young men, and hence that its interpretation allowing women to receive scholarships from the fund was consistent with his wishes. This
seems little short of preposterous, but one can see that the court
thought that the law clearly favors giving effect to the intent of the
testator, and saw no other unproblematic way to reach the desired result. On its face this appears to be a case in which the most reasonable theory to be extracted from settled law would yield a decision
rule restricting these scholarships to males. But it might be that a
more candid court could have justified the decision actually reached
partly on the basis of social and legal changes that have taken place
since the will was drafted in 1963: it might, that is, have tried to
claim that this is an area in which the law is in flux, and that the
theory that best justifies that law is changing from what it would have
been a decade or two ago.

On the other hand, it might be that there is no way ultimately to
justify the decision without making a *moral* case for the underlying du-
ties being presupposed. It is this moral appeal that is not required
in justification in science and Dworkin obviously thinks this is a difference that *makes* a difference. Clearly he wants to insist on the
different roles played by coherence in the legal and scientific models:
in law it is fundamentally a moral requirement of equality or justice
whereas in science it is a more purely intellectual demand. But this,
I suspect, is really to say that the differences between the two have
more to do with content than with form, and that view I find quite acceptable.

Dworkin's theory has been referred to as the "Third theory of
law"[57] in that it is meant as an alternative both to legal positivism

and to natural law doctrines. What makes his work particularly slip-
pery is that it contains elements of both those positions, and they
are hard to reconcile. It is like a natural law theory in that legal
propositions are true or false and rights are pre-existing; but it is
like a positivist theory in that these rights do not represent transcendent realities but are rather constructed from the fabric of legal
decisions. The underlying assumptions appear to include somewhat simple versions of scientific realism, moral objectivism, and the doctrine that rights can at least in principle be balanced in some objective way, perhaps even along a linear scale. Much of this is un-
exceptionable. My quarrel with him is not about his claim that judicial decision rests on a coherence theory of justification, for I
think he is right about that, at least as a piece of idealized legal
methodology. Where I think he is mistaken is about the appropriate
notion of coherence, both for the law and for science. By insisting
that the former is theory-centred and the latter fact-centred he per-
petuates a false dichotomy that has informed most thinking about science until very recently. This is the idea that scientific facts have
some sort of independent existence in a world in which they simply
await discovery, and once we have collected enough of them we can develop some sort of theory. But this picture is both false for science
and out of step with Dworkin's own conception of coherence. A better
picture would, I suggest, be a modification of his constructive model
to achieve a greater balance between the demands of fact or moral intuition on the one hand and theory on the other, with no *a priori* pro-
nouncement about which wins when there is a conflict. Such a view
would allow us to see the deep similarities between justification in
the two realms without obscuring the differences that stem from the
ineradicable presence of a moral dimension in legal theorizing.

FOOTNOTES

1 The literature, of course, abounds with discussions of verification, falsification and demarcation. Representative examples of
   Popper's views can be found in his *Conjectures and Refutations*,
   Basic Books, 1962, chapters 1 and 11. A particularly lucid version of the logical positivist perspective is A. J. Ayer's *Language, Truth and Logic*, Dover Publications, 1946. It is perhaps
   worth remarking here that I am concerned in this paper mainly
   with problems whose focus has traditionally been in the context
   of justification, rather than in the context of discovery, insofar
   as that distinction can be maintained.

2 For recent discussions of inference to the best explanation see
   M. Hanen, "Confirmation, Explanation and Acceptance", in K. Lehrer
   (ed.), *Analysis and Metaphysics*, D. Reidel Publishing Company, 1975;
   G. Harman, *Thought*, Princeton University Press, 1973; and P. Thagard, "The Best Explanation: Criteria for Theory Choice", *Journal
   of Philosophy*, LXXV, 2 (February 1978). The idea, of course, derives from C. S. Peirce.

3 H. J. Berman, "The Origins of Western Legal Science", *Harvard Law
   Review*, 90, 5 (March 1977), p. 931.

4 *Ibid.*, p. 931.

5 *Ibid.*, p. 932.

6 *Ibid.*, p. 932.

7 I discuss the relation between law and literary criticism below.

8 See his *Taking Rights Seriously*, (hereinafter *TRS*), Harvard University Press, 1977, especially chapter 4.

9 This distinction appears in Dworkin's earlier papers, such as "The
   Model of Rules", which appeared in 1967 and is reprinted as chapter
   2 of *TRS*. But it was somewhat modified in "Hard Cases", which
   first appeared in 1975, and is chapter 4 of *TRS*.

10 *TRS*, p. 105.

11 *TRS*, p. 116.

12 *TRS*, p. 119.

13 J. Rawls, *A Theory of Justice*, (hereinafter *ATOJ*), Harvard Univer-
    sity Press, 1971, p. 20. Rawls actually distinguishes between wide
    and narrow reflective equilibrium in a later paper, "The Indepen-
    dence of Moral Theory", *Proceedings and Addresses of the American
    Philosophical Association*, Vol. XLVIII (1974-75), but this distinc-
    tion is not important for our purposes.

14 *ATOJ*, p. 20.

15   N. Goodman, *Fact, Fiction and Forecast*, 3rd ed., Hackett Publishing Company, 1979, p. 64.

16   J. Feinberg, *Social Philosophy*, Prentice-Hall, 1973, p. 34.

17   *ATOJ*, p. 49.

18   *ATOJ*, p. 47.

19   *ATOJ*, p. 49.

20   *ATOJ*, p. 49.

21   This paper appears as chapter 6 of *TRS*, though it was first published in 1973.

22   *TRS*, pp. 279-80, emphasis added.

23   R. Dworkin, "Seven Critics", in *Georgia Law Review*, Vol. II (1977), p. 1249, emphasis added.

24   *Ibid.*, pp. 1202-3.

25   *TRS*, p. 160.

26   *TRS*, p. 160.

27   *TRS*, p. 160.

28   *TRS*, p. 160.

29   *TRS*, p. 161.

30   *TRS*, pp. 164-5. The reference to Quine is to "Two Dogmas of Empiricism", in W. V. Quine, *From a Logical Point of View*, 2nd ed., Harvard University Press, 1964.

31   *Brown v. Board of Education of Topeka, Shawnee County, Kansas*, 347 U.S. 483, 74 S.Ct. 686, 98 L.Ed. 873.

32   See, for example, Robert E. Keeton and Jeffrey O'Connell, *Basic Protection for the Traffic Victim: A Blueprint for Reforming Automobile Insurance*, Little, Brown and Company, 1966.

33   See T. S. Kuhn, *The Structure of Scientific Revolutions*, 2nd ed., University of Chicago Press, 1970, especially chapters VI-VIII.

34   C. M. Turbayne, *The Myth of Metaphor*, Yale University Press, 1962, p. 174. Turbayne, of course, wants to claim that the anomaly of the Barrovian case can be resolved by shifting from the geometrical theory to the linguistic theory that he takes Berkeley to have been advancing in *An Essay Toward A New Theory of Vision*.

35   *Ibid.*, p. 175.

36   P. Feyerabend, in his *Against Method*, Verso, 1978, p. 60, has this to say about the example of Barrow:

> Barrow *mentions* the qualitative difficulties, and he
> *says* that he will retain the theory nevertheless.
> This is not the usual procedure. The usual procedure
> is to forget the difficulties, never to talk about
> them, and to proceed as if the theory were without
> fault. This attitude is very common today.

Feyerabend's book is in part an extended argument, using Galileo as
a case study, that precisely the kind of ignoring of data that
Rawls and Dworkin countenance only for moral theory does and should
go on in science. I argue that these views are exaggerated, though
they do provide a useful antidote to a too rigid conception of
methodology.

37 M. Polanyi, *Personal Knowledge*, The University of Chicago Press,
    corrected edition, 1962, pp. 12-13. Actually there is good reason
    to think that Polanyi's account is an oversimplification, for the
    adherents of Special Relativity apparently did not simply ignore
    negative evidence; rather, they "surmised...that the conditions
    under which Miller obtained his results were different from what he
    had supposed them to be." For a thorough and helpful discussion of
    this issue see A. Grünbaum, *Philosophical Problems of Space and
    Time*, Second, enlarged edition, D. Reidel Publishing Company, 1973,
    pp. 379-386, from which the above quotation is taken. What we seem
    to have here is another instance of the kind of theorizing I have
    suggested goes on in science *and* morals, where anomalies are sup-
    posed to have some explanation, even if we can't at the moment see
    just what it is.

38 L. Laudan, "Two Dogmas of Methodology", *Philosophy of Science*, 43,
    4 (December 1976), pp. 589-90, 591. Laudan attributes the second
    example to R. Laudan. Related points are made by P. Feyerabend,
    op. cit., and by A. Grünbaum in "Can a Theory Answer more Questions
    than One of Its Rivals?" *British Journal for the Philosophy of
    Science*, 27 (1976), pp. 1-23.

39 *TRS*, p. 161. I do not mean to claim here, of course, that there is
    no tie to observation in science, or that anything goes. Rather,
    it is that the observational tie is complex, and may not be differ-
    ent in principle from the moral sphere, where we have ties to the
    realities of human nature and social organization.

40 For a discussion of the lottery paradox (first formulated by H.
    Kyburg) and its significance for a theory of acceptance, see M.
    Hanen, "Confirmation, Explanation and Acceptance".

41 *Ibid.* See also G. Harman, *Thought*; Bas C. van Fraassen, "The Prag-
    matics of Explanation", *American Philosophical Quarterly*, XIV, 2
    (April 1977); and Paul R. Thagard, "The Best Explanation: Criteria
    for Theory Choice".

42 See I. Scheffler, "On Justification and Commitment", *Journal of
    Philosophy*, Vol. L, (1954).

43 *TRS*, p. 283.

44 See, for example, H. L. A. Hart, *The Concept of Law*, Oxford University Press, 1961, especially chapter VII. It is, in fact, Hart's theory that Dworkin is criticizing when he argues that judges do not have discretion, and that there is always a single right answer in any civil case.

45 R. Dworkin, "No Right Answer?" in P. M. S. Hacker and J. Raz, *Law, Morality, and Society*, Oxford University Press, 1977.

46 *Ibid.*, p. 73.

47 *Ibid.*, pp. 73-4.

48 *Ibid.*, p. 75

49 *Priestman v. Colangelo, Shynall and Smythson*, [1959] S.C.R. 615, 30 C.R. 209, 124 C.C.C. 1, 19 D.L.R. (2d) 1; *Beim v. Goyer*, [1965], S.C.R. 638, 57 D.L.R. (2d).

50 P. Weiler, *In the Last Resort*, Carswell/Methuen, 1974, p. 75.

51 *Ibid.*

52 *Priestman v. Colangelo*, [1959] S.C.R. 615 at 623, 624-25.

53 *Beim v. Goyer*, [1965], 57 D.L.R. (2d) at 255.

54 *Ibid.*

55 For a thoroughly radical view about the correspondence between the world and our pictures and descriptions of it, see N. Goodman, *Ways of Worldmaking*, Hackett Publishing Company, 1978.

56 *Ebitz v. Pioneer National Bank*, Mass. 361 N.E. 2e 225.

57 See J. L. Mackie, "The Third Theory of Law", *Philosophy and Public Affairs*, vol. 7, no. 1 (1977).

II. THE IMPACT OF PSEUDO-SCIENCE ON THE DEVELOPMENT OF SCIENCE

The historical relationship between science and pseudo-science has hardly been one of strict quarantine. On the contrary, the relationship has often been one of mutual interpenetration. Pseudo-science draws its notion of science from the more respectable branches of scientific pursuit; indeed it implicitly assumes the authoritative status of scientific forms of knowledge. Nonetheless, the roots of conceptual innovations within the accepted body of science sometimes lie far beyond traditional disciplinary boundaries and may indeed arise from areas called pseudo-scientific. And examination of a thinker's attitudes toward the pseudo-sciences may shed light on his attitude toward methodology and the proper scope of scientific study. The papers in this section illustrate these points.

Westfall's paper on the influence of alchemy on Newton's science directly challenges the notion of linear progress in the growth of scientific thought, arguing that Newton's concept of attractive force - surely the most profoundly important new idea in the *Principia* - had its source in the alchemical studies that occupied a great portion of Newton's intellectual life. That alchemy, by almost anyone's criteria a candidate for the label "pseudo-science," was the source of a scientific concept of such profound significance, challenges the progressive interpretation of science for it raises the question of whether Newton, or anyone *in medias res*, can determine *a priori* which lines of research will be fruitful and which will not.

Newton's immersion in alchemy may help further our understanding of the relationship between science and pseudo-science. The fact that Newton kept his alchemical studies secret, even though they occupied a far greater proportion of his life's work than the scientific studies which he made public, speaks to the strong possibility that in his mind, at least, there was a clear line of demarcation between legitimate natural philosophy and other pursuits. This supposition is reinforced by the distinction that separated the main contents of the *Principia* and the *Opticks* from the more speculative matters treated in the various scholia and "Queries" appended to these volumes. Perhaps our presently intuitive distinction between science and pseudo-science stems from the influence Newton's implicit distinction has had on our own thinking.

Levere's paper shows how Coleridge's concern with the pseudo-sciences of Phrenology and Mesmerism, as well as his interest in science itself, can be understood within the context of his life-long "striving after the unrealisable goal of the self-integration of all knowledge." The search for a unified view of knowledge and nature led Coleridge to develop views on scientific method, the structure of nature, and man's place in nature. Although his presuppositions were those of Romantic Idealism - *Naturphilosophie* - a view scorned by hard-nosed empiricists, Coleridge's approach was often clear-sighted and hard-headed. He was attracted to the new disciplines of Anthropology, Phrenology, and Mesmerism because he believed that they would form a scientific approach to man that would fit in with his broader views of nature in general. Later science may have rejected Phrenology and Mesmerism as pseudo-sciences, but Coleridge turned to them in search of an explanation of the observed facts of Anthropology. The structure of his thought might find its source in metaphysics, but he nonetheless accepted empirical evidence as the ultimate test of a theory.

McKillop's paper on the enigmatic character Daniel Wilson shows how religious and moral assumptions can be a determining influence on science and theories of nature. Wilson's response to Darwin, influential in Canadian intellectual history, sprang more from his defense of Christian belief than from his views of scientific method. The role of values in the evaluation of scientific theories may be unusually explicit in Wilson's case, but the practice has been so widespread in the history of science as to call for a reconsideration of the myth of value-free scientific objectivity.

THE INFLUENCE OF ALCHEMY ON NEWTON

The influence of alchemy on Newton was not a static condition which can be analysed as though it happened once and for all and underwent no change. The topic demands a chronological approach. Like the tide Newton's concern with alchemy rose and fell, and like the tide it deposited some of its burden, which remained on the shore altering its contour. We can understand the phenomenon only if we follow it through the entire cycle.

Let me torture my simile for one last analogy. We can only identify what the tide has brought in if we know what the shore was like before. Newton's interest in alchemy was not his earliest intellectual activity. It began against a well defined background. Unless we survey the background in detail - the shore at low tide - we cannot hope to assess the changes the tide wrought. Fortunately Newton was good enough to preserve a record of the background, apparently aware that in the 20th century a strange breed of academics would arise, a race much given to organizing symposia on sundry arcane topics, who might find in his manuscripts food for talk and even perhaps for thought. With rare forebearance he refrained from destroying the miscellaneous papers which few men keep. As a result, we possess even the records of his undergraduate studies; I turn to them to sketch the setting for my story.

More than fifty years after his undergraduate career, in a conversation with an itinerant Venetian noble, the Abbé Antonio-Scinella Conti, who was resident for a time in London and became his confidant, Newton confessed that he had been converted to Cartesianism in his youth.[1] We can in fact follow his conversion in reading notes from his undergraduate days. From all appearances the official curriculum had nothing to do with the process. The universities were the most reactionary institutions of 17th-century society. By the early 60's a revolution in thought, perhaps the most profound revolution the entire intellectual history of European civilization has witnessed, had been in progress for more than two generations. It had not yet seriously penetrated the walls of the universities which continued to feed the students on the bones of Aristotelianism bleached dry by the suns of four centuries. Newton's reading notes reveal that he too, in his

turn, was made to plunge into the same stale waters, reading in Aristotelian logic, rhetoric, ethics, physics, and philosophy in general as
they were served up in the textbooks of the Stagirite's 17th-century
epigoni. When I said that the recent revolution in European thought
had not penetrated the walls of the universities, I meant officially.
We have independent testimony that Descartes, for example, was being
read unofficially in Cambridge, that his philosophy was spreading underground through what one witness called the brisker part of university,
and Newton's notebook establishes that it made its ways to the chamber
in Trinity College that he shared with John Wickins.[2] He left the
notes he was taking from various peripatetic textbooks incomplete, simply abandoned the established curriculum on which his own academic future might hinge, and plunged into an extended program of reading in
the new philosophy. If we allow the evidence in his notebook to guide
us, we may choose to question his assertion to Conti that he became a
Cartesian. Rather, we may find it necessary to place a broad interpretation on the word "Cartesian" so that it means, not the specific philosophy of Descartes, but a somewhat less precise mechanical philosophy.
Certainly Newton read Descartes. He did more than read him. He devoured him, and nearly every page of the notebook in question contains
evidence of the feast. Descartes was not the only author on his menu,
however. He read as well Galileo's *Dialogue*, Kenelm Digby, Joseph
Glanvill, Henry More, Robert Boyle, and Walter Charleton's English epi-
tome of Pierre Gassendi's rendition of atomic philosophy. The later
especially caught Newton's attention, and the record of the notebook
strongly suggests that already as an undergraduate he found the Gassen-
dist atomic philosophy more attractive than the Cartesian. For my purposes today, the distinction does not matter. Let me say merely that
before he took his Bachelor's degree in 1665, Newton had embraced the
mechanical philosophy of nature. I take this to be what he had in mind
when he told Conti that he became a Cartesian.

The mechanical philosophy of nature, largely the creation of the
generation preceding Newton, was an attempt to separate the psychic and
the spiritual from the physical and material and to establish, within
the realm of natural philosophy, the autonomy of material causation.
Descartes' *Meditations* provided the cornerstone of its metaphysical
foundation. In the sixth and last meditation, using the results of
the previous five, Descartes demonstrated to his own satisfaction the
existence of the material realm which he had called into question at

the very beginning of his process of systematic doubt in the first meditation. Though it is necessary that the physical world exist, he continued, there is no corresponding necessity that it be in any way similar to the world our senses depict.[3] No other statement in the century better summarized the program of the mechanical philosophy which
dedicated itself to the proposition that the underlying reality of nature is utterly different from its surface appearance. On the surface
nature presents itself to us, as virtually the unanimous tradition of
Western philosophy had accepted it, in organic terms. In reality it is
a complex machine; the very concept of life is an illusion. So is much
else. With the exception of extension, figure, and bulk (a list which
varied somewhat from man to man), all of the properties attributed to
bodies in the old philosophy were also stamped as illusions, the projection of our subjective sensations onto the external world. Nature
is only a complex machine. Reality consists of particles of matter in
motion, and particles of matter in motion produce all of the phenomena
we observe. There is no necessity and indeed no reason to suppose that
nature is in any way similar to the world our senses depict.

For my purposes it is important that the mechanical philosophy also banished from existence another denizen of some previous philoso-
phies - attractions of any kind. No scorn was too great to heap upon
such notions. From one end of the century to the other, the idea of
attractions, the action of one body upon another with which it is not
in contact, was anathema to the dominant school of natural philosophy.
Galileo could not sufficiently express his amazement that Kepler had
been willing to entertain the puerile notion, as he called it, that the
moon causes the tides by acting upon the waters of the sea.[4] In the
90's, Huygens and Leibniz found similar ideas just as absurd for the
same reasons.[5] To speak of an attraction whenever one body was seen to
approach another was to philosophize on the same plane with Molière's
doctor who explained the power of opium to cause sleep by a dormative
virtue it contained. "Bene, bene, bene," cried his examiners; "bene
respondere," and Molière's ridicule was that of the entire century.[6]
An attraction was an occult virtue, and "occult virtue" was the mechanical philosophy's ultimate term of opprobrium.

Two specific phenomena serve as examples. To the characteristic
natural philosophy of the 16th century, what has been called Renaissance Naturalism, magnetism was the ready exemplar of the vast range
of mysterious forces with which it conceived the universe to be

populated. Mechanical philosophy viewed magnetism as something it must explain away if its program were to claim validity. In his *Rules for the Direction of the Mind*, Descartes treated it almost as a test case. When the ordinary person confronts magnetism, he said, he immediately assumes that it is utterly unlike anything he is familiar with and hence grasps at whatever notion is most difficult. One ought rather to review things that are familiar and to find the points wherein they agree with the properties of magnets.[7] In the *Principles of Philosophy* Descartes took his own advice to heart. He listed in detail the known phenomena of magnetism, and he imagined a mechanism whereby the direct contact of moving particles accounted for everything. He described how the turning of the vortex on its axis produces tiny screw shaped particles and similar pores in terrestrial material to receive the particles. He even pointed out how the motion of particles along the two opposite poles of the vortex must produce both left hand and right hand screws, corresponding to the two poles of the magnet. What is the reality behind the apparent attractions of the magnet? Streams of invisible particles which move ferrous bodies by direct contact.[8] Descartes' explanation of magnetism is one of the wildest flights of mechanistic imagination from the whole 17th century, and others of his age found it so. I am not aware of anyone who followed him on this topic, but every mechanical philosopher, far from rejecting his program, invented his own alternative mechanism consisting of invisible streams or of threads with grappling hooks or the like. Isaac Newton, the young mechanical philosopher in Trinity College, also turned his mind to the problem. He imagined two streams of particles which, in passing through iron, acquire distinct "odors" which make them sociable, one stream to iron and loadstone, the other to the aether that fills their pores.[9] Perhaps I was wrong to label Descartes' explanation the wildest flight of mechanistic imagination; it might be hard to judge Newton's any less wild. Fascinated by the unceasing flux of matter that such mechanisms seemed to present, he thought of tapping into it with some machine to realize a perpetual motion. In his undergraduate notebook he sketched in two different devices, mounted on axles and turned perpetually by the magnetic stream.[10]

Gravity offered a similar challenge. Aristotle had ascribed the tendency of bodies to fall to their nature as heavy bodies. More recent philosophers had attributed the fall of bodies to an attraction. Descartes derived it instead from the mechanical necessities of the

vortex. Every body moving in a circle strives to recede from the center. In a plenum, some can recede only if others move in the opposite
direction. What is gravity? A deficiency of centrifugal force on the
part of certain bodies which other bodies drive toward the center.[11]
Once again, alternative mechanical explanations of gravity presented
themselves, usually in terms of the descent toward the earth of some
invisible matter which bears down bodies in its path. Newton's student
notebook contains passages about the gravitating matter. In connection
with it he thought once more of perpetual motion and imagined the possibility of a gravitational shield such that a wheel hung on a horizon-
tal axis with half of it extending beyond the shield would turn perpetually.[12]

In 1675, in connection with a paper of observations of the colors
in thin transparent films, Newton sent an *Hypothesis of Light* to the
Royal Society. He had noted, he said, that the heads of certain vir-
tuosi ran to hypotheses, and he offered his as a possible aid to help
them understand his theory of colors. The *Hypothesis of Light* dealt
with more than light, however; it tied together various speculations
into a total, mechanistic, system of nature. Central to it was a universal aether which pervades all of space. Newton explained the reflection, refraction, and diffraction of light by a simple mechanism in
the aether. Standing rarer in solid bodies than in free space, for example, it refracts light by the differential pressure at a refracting
surface which pushes the corpuscles of light and changes their direction. Wave motions in the aether explained the phenomena of thin
films. The aether also explained much beyond optics. Its condensation
in the earth, for example, produces a steady downward flow and causes
gravity. He described a simple experiment in static electricity in
which the rubbing of a piece of glass, as he explained, rarefies the
aether condensed in it, making the aether stream out and move tiny bits
of paper. It is true that the *Hypothesis of Light* contained other elements which do not lend themselves to easy summary as a mechanical
philosophy of nature. Nevertheless, the mechanistic elements in the
essay were no less real than the other aspects, and for the moment they
are the part I wish to emphasize.[13]

Newton composed the *Hypothesis of Light* in 1675. Twelve years
later, as I will not need to remind anyone here, he published the *Principia*, in which a rather different set of concepts displayed themselves.
Shortly before the *Principia* appeared, Fatio de Duillier arrived in

England after a stay in the Netherlands. In June 1687 he wrote to
Huygens telling him about the book on the system of the world which
Newton would soon bring out, a book his English friends were telling
him would revolutionize natural philosophy. The English, he indicated,
found him too Cartesian, but he wished that Newton had consulted Huygens on the principle of attraction which his book proposed. Huygens
replied that he did not care whether Newton was a Cartesian or not "as
long as he doesn't serve us up conjectures such as attractions."[14]
Serve up a theory of attraction Newton did, more of a theory of attraction than Huygens could have imagined had he let his mind run freely
over all that he considered most misguided. Having proposed a planetary dynamics based on the concept of centripetal attraction, Newton
went on to argue that every particle of matter in the universe attracts
every other particle, a theory which could not have flouted mechanical
sensibilities more openly.

   The *Principia* carried several disclaimers, of course. He treated
centripetal forces as attractions in Section XI, he said, "though perhaps in a physical strictness they may more truly be called impulses.
But these Propositions are to be considered as purely mathematical;
and therefore, laying aside all physical considerations, I make use of
a familiar way of speaking, to make myself the more easily understood by
a mathematical reader." And again: "I here use the word *attraction* in
general for any endeavor whatever, made by bodies to approach to each
other, whether that endeavor arise from the action of the bodies themselves, as tending to each other or agitating each other by spirits
emitted; or whether it arises from the action of the ether or of the
air, or of any medium whatever, whether corporeal or incorporeal, in
any manner impelling bodies placed therein towards each other. In the
same general sense I use the word *impulse*, not defining in this treatise the species or physical qualities of forces, but investigating the
quantities and mathematical proportions of them..."[15] Later on, in the
second English edition of the *Opticks* he inserted a series of Queries
which proposed the existence of an aether which, among other things, he
supposed to cause gravity, not by streaming toward the earth and sun,
but by a constant differential pressure which increases steadily with
distance from bodies.[16]

   For all the disclaimers however, and despite the Queries in the
Opticks, it is impossible to imagine a mechanism adequate to the job of
universal gravitation. As he prepared the text of the second edition

for the press, Roger Cotes tried to raise this problem with Newton in
connection with the third law. As far as the attraction of the sun on
a planet was concerned, Newton might argue that he merely used the word
"attraction" without making any claim about its ontological status.
One could imagine an aetherial mechanism to explain it. By the third
law, however, and according to the necessities of Newton's system, the
planet also attracts the sun. What aetherial mechanism would explain
this?[17] Although Newton swept the objection aside, it made a valid
point. Other aspects of the *Principia* reinforced it. If one is not
too concerned with the niceties of the lunar orbit, he can imagine
some aetherial device to explain the moon's centripetal "attraction"
toward the earth. Newton also devoted some attention, however, to the
moon's attraction on the earth, on the one hand raising the waters of
the sea in the tides, on the other hand moving the bulge of matter
about the equator and causing the earth to precess. What did Query 21
have to say about these phenomena? What did it have to say about Proposition VII of Book III of the *Principia*, "That there is a power of
gravity pertaining to all bodies, proportional to the several quantities of matter which they contain," such that every particle of matter
in the universe attracts every other particle? The fact is, Newton's
disclaimers did not make sense. The *Principia* was about attractions,
and the concept of attractions flew in the face of the orthodoxy accepted by the entire scientific community of his day. If the word
"revolution" will bear one more use, surely this was a scientific revolution.

Here then is the question to which I address myself. The concept
of attractions, though certainly not unknown before, represented the intrusion of a new and alien idea into the accepted body of natural philosophy. I wish to insist on the conceptual issue. I cannot myself
find any way to extract the idea, say, from the mathematical necessities of the situation. Newton was not the only sophisticated mathematician of the day. Christiaan Huygens, Gottfried Wilhelm Leibniz, and
Johann Bernoulli were, after Newton, the leading mathematicians of
Europe at the time, in the opinion of historians rather considerable
mathematicians all. Even with the *Principia* laid open before them,
they were unable to find that its mathematical demonstrations led to
the idea of attractions, and to their deaths not one of them accepted
it. We are dealing with a conceptual issue, not a mathematical one. I
pose the question: can we find the source from which Newton drew it?

I pose the further, more specific question: do we have in the idea of
attractions an enduring influence of alchemy on Newton's scientific
thought?

Alchemical study did not go back to Newton's undergraduate days.
In no real sense did chemical study. Robert Boyle was one of the authors he learned to know, however, and if he read him first as a mechanical philosopher, it was not long before he was reading him as a
chemist. Sometime in the 60's, after his undergraduate days had ended
if we can judge by the hand, he composed a glossary of chemical terms,
such things as "Acid salts spirits and juices," "Aqua fortis," and
"Cementation." No odor of alchemy clings to the material that he entered under these headings; it smells rather of sober chemistry.[18] His
recipe for making phosphorus also smelled of sober chemistry, and of
something else besides. "Take of urine," it began, "one barrel."[19] It
is true that his glossary included headings such as "Alkahest" and
"Projection," but it is equally true that he did not enter anything under them. By the late 60's his reading was changing. His accounts for
1669 show the purchase of the huge six volume *Theatrum chemicum*, the
greatest collection of alchemical texts ever published, during a visit
to London, and notes from it as well as from other alchemical writers
such as Basil Valentine, Sendivogius, Eirenaeus Philalethes, and
Michael Maier, notes written in the hand of the late 60's, survive.[20]
On the trip to London he also purchased chemicals and two furnaces, and
one of his notebooks records experiments apparently from about this
time.[21] It is important to note the order of his progress. Newton did
not stumble into alchemy as a youth, recognize the emptiness of its
promise, and turn instead to rational chemistry. Rather he started
with rational chemistry (in so far as a distinction between "rational"
chemistry and alchemy is valid for the 17th century) and worked his way
into alchemy. The man who purchased the *Theatrum chemicum* in 1669 had
already invented the fluxional calculus, discovered the heterogeneity
of light, and entertained rudimentary hints of the law of universal
gravitation. For the following twenty-five years, while he scarcely
touched optics except to present work done earlier, while he turned to
mathematics less and less frequently, while he ignored what he had done
in mechanics until a chance visit from Halley triggered an intense investigation for two and a half years in the middle 80's, for the following twenty-five years, I say, alchemy was his most consistent scientific activity.

Let us be clear that we are talking about an extensive application
of effort to which an extensive body of surviving papers testify. I
have devoted some time to a reliable quantitative measure of these papers. They appear to contain well over a million words in Newton's
hand. An activity that left behind a record that large was clearly
more than an incidental diversion.

The papers fall into various categories. Many of them are reading
notes. It is frequently asserted that all of them are reading notes,
so that from them one cannot conclude anything with assurance about
Newton's own attitude toward the subject. This is simply not true.
Meanwhile the reading notes themselves cannot be dismissed that lightly
since they bear testimony to study of the alchemical tradition which
was both intensive and extensive. Eighty-seven closely written pages,
for example, explored the works of Michael Maier.[22] Newton was always
concerned to compare one alchemist with another in order to winnow the
grain from the chaff. Early in the 80's, he began to compile an *Index
chemicus*, as he called it, which would help him to control the body of
information he was compiling. The *Index chemicus* advanced through
three successive stages. It began with 115 headings entered on two
sheets folded twice to give him eight folios in all, and under the
headings he entered page references to places where he could find those
topics discussed. He also squeezed new headings in until he exhausted
the space available and started anew on twenty-four folios, now with
251 headings. The second version expanded as the first had done,
reaching 714 headings before it could hold no more. It also began to
change its character. Initially the *Index chemicus* appears to have
been a typical Newtonian device to organize and command his information. He invariably began every new study with something similar.
Some of the entries began to take on the form of short essays, however,
and as he started to expand the *Index* anew, Newton devoted some effort
to drafts of the essays on separate sheets. He started a new version,
abandoned it before he completed the letter A, and treated its material
as further drafts. The second version of the *Index* contained a few
references to Mundanus, as Newton called Edmund Dickinson's letter to
Theodore Mundanus published in 1686, references always squeezed onto
the ends of lines. While it is impossible to know exactly when he
read Mundanus, it is reasonable to date the beginning of the final version of the *Index* about 1686 or 87. This final version, final in the
sense that no further one ever replaced it, though not final in the

sense that it was ever completed, stretched beyond one hundred pages
and contained 879 headings. I analyzed the content of the 46 largest
entries, which together fill about 42 pages and thus constitute a sizable proportion of the whole. I counted 1,975 separate page references
to 141 separate treatises and 104 different authors. By extrapolation,
the entire *Index* must contain about 5,000 page references, undoubtedly
with several more authors not represented in the longest entries. The
Index reveals that Newton had consumed the contents of all of the major
collections of alchemical works, *Theatrum chemicum*, *Artis auriferae*,
Musaeum hermeticum, *Theatrum chemicum Britannicum*, and *Aurifontina
chymica*. He had digested as well the various works of all the major
figures in the long history of alchemy - men such as Arnold of Villa-
nova, Raymond Lull, Nicholas Flamel, Johann Grasshoff, Michael Maier,
and the pseudonymous Eirenaeus Philalethes, an English alchemist still
alive when Newton took up the Art and the master of it who influenced
him the most.[23] Newton collected as he read. More than ten percent of
his library at the time of his death consisted of alchemical works;
since he hardly collected during his final twenty-five years, the proportion at one time must have been far greater.[24] Although I am not
really in a position to judge, I am willing to venture the opinion that
alchemy has never had a student more widely and deeply versed in its
sources.

One peculiar feature of his alchemical studies deserves mention.
Among his papers is a large sheaf of alchemical essays copied in several hands, almost all essays which have never been published. In the
sheaf are one paragraph and several corrections in Newton's hand of the
mid or late 60's, and elsewhere notes that he made from the collection
survive.[25] The notes suggest to me that he borrowed this collection
from someone to study and for whatever reason failed to return it. He
also possessed a number of copies, in his own hand, of other alchemical
tracts that have never been published. His writing in these copies
indicates that they stem, not from any one period, but from the whole
extent of his career in alchemy. His copy of one states that "Mr. F"
(in the Cambridge context, probably Magister F, an academic colleague)
gave it to him in 1675. One Ezekiel Foxcroft, a member of the alchemical circle that had gathered initially around Samuel Hartlib,
lived in King's College at this time; Newton later studied his translation of the Rosicrucian *Chymical Wedding* and referred to it as
Mr. F's work. The manuscripts of Eirenaeus Philalethes circulated
originally through the Hartlib circle to which Foxcroft belonged;

Newton had access to some of them about ten years before their publication. In 1683, he received a letter solely concerned with the Art from one Fran. Meheux, and in the 90's, scarcely a month before he accepted a position which put him in charge of His Majesty's coinage in gold and silver, an unnamed alchemist looked him up in Cambridge to discourse about the Work.[26] I see no way to interpret these various facts except to conclude that Newton was in touch with clandestine alchemical circles from which he received unpublished alchemical literature. It seems very likely to me that he also fed material into the network. His name would not have appeared on such papers. Until very recently, those familiar with alchemical materials have not exploited Newton's manuscripts and have not been familiar with his handwriting. I shall be surprised if we do not begin soon to discover hitherto unknown Newtonian writings on the Art.

It has long been known that Newton left a large batch of alchemical papers behind. As I mentioned, it has most frequently been asserted, incorrectly, that they consist entirely of reading notes. Quite a few of them are Newton's own compositions. Rather early in his studies he drafted a somewhat disjointed paper, usually called *The Vegetation of Metals*, which contains a long essay on alchemical themes. In the late 70's he drew upon his own experimentation in a paper called *The Key*, which Mrs. Dobbs has explicated with great success, and about the same time another piece also closely related to experience in the laboratory on *The Separation of Elements*. He wrote a commentary on the *Tabula smaragdina*.[27] He was much concerned to draw the alchemical authorities, which he knew so well, together into one consistent statement of the Work. There is a sheet with seventeen titles which sound like the titles of chapters, and elsewhere drafts of chapters, filled with material drawn from alchemical writers, with titles which correspond to those on the list.[28] A paper called *The Regimen* sums up the Work in seven "Aphorisms," the process, he stated, as he found it in "ye work of the best Authors, Hermes, Turba, Morien, Artephius, Abraham ye Jew and Flammel, Scala, Ripley, Maier, the great Rosary, Charnock, Trevisan. Philaletha. Despagnet." Four long pages supported the aphorisms with citations from these authors.[29] Another somewhat later compilation started with the title *Decoctio* but shifted in a later version to the name of the earlier paper, *The Regimen*.[30] *The Method of ye work*, originally a comparison of Didier's work with that of eleven other authors, went through a second stage which concerned

itself principally with Ripley, and ended as a paper, which exists in
two drafts, *Praxis*, arguably the most important alchemical paper which
Newton ever wrote and one which explicitly claims to have achieved multiplication.[31] The most extensive of his compilations, which carried
no title, was divided into a number of chapters or "opera." We have
six drafts of one of the opera, four of another, at least two of three
more.[32] In many of these papers, especially the last one, most of the
content consists of citations from others. Newton must have known his
favorite authors almost by heart. In some of the papers he left empty
parentheses after citations where he could later fill in page numbers
to the passages in question. Nevertheless, these were his own compilations, testifying to his conviction that all true alchemists pursued
the same Great Work. The papers were the result of extensive study and
thought. There is no way to dismiss them as reading notes.

Finally, Newton pursued alchemy in the laboratory as well. Notes
of his own experiments are among his earliest chemical papers. When
Humphrey Newton was with him, from 1683 until 1688, he found him exper-
imenting in the garden laboratory outside his chamber in Trinity; Humphrey's most vivid recollection of those years centered on the laboratory.[33] Recently Mrs. Dobbs has correlated Newton's early experiments
with his alchemical papers and proved to my satisfaction that we can
only call the experimentation alchemical. Experimental records continue through the 80's and into the 90's. They still await final explication, but the interjection among sober laboratory records of comments such as "I saw Sophic Sal Ammoniac," "I made Jupiter fly on the
wings of the eagle," serve to convince me at least that this experimentation too, like the earlier, was alchemical.[34] The last dated experimental notes come from February 1696, less than a month before the
Lords Commissioners of the Treasury appointed Newton Warden of the
Mint.[35] One can only surmise that they had not heard what was going
on in the garden next to the great gate of Trinity.

Newton's alchemical papers have been a source of anguish to many
scientists to whom they seem the negation of everything to which they
would assign the name science and thus of everything for which Newton
stands in their eyes. Newton's great biographer in the 19th century,
Sir David Brewster, could scarcely bring himself to confess that his
hero had not only copied "the most contemptible alchemical poetry"
but had even annotated "the obvious production of a fool and a knave."
Indeed the only solace Brewster could extract from the situation, and

he clearly found it little enough, was the fact that Leibniz had also dabbled in the Art.[36] Perhaps we are no longer so strict in our notions of proper scientific behavior. At least in recent years an increasing number have begun to look seriously at the papers. Their interpretation will probably remain a matter of contention, but a few facts have at least become clear. The papers are undoubtedly authentic. As I indicated, they are quite extensive. They are the products neither of Newton's youth nor of his age, but of the middle years of his life when he was at the height of his powers, the years when he completed both of the books on which his enduring reputation in science rests. To me at least it appears obvious that any activity which succeeded in absorbing that much of his time and energy must have held great meaning for him. The chronology involved, spanning the time when he arrived at the concept of attraction with which he revolutionized natural philosophy, seems to invite us to explore the possibility that the alchemical tradition could have offered a source from which he drew, not the idea, but stimulus toward it.

Compared to the mechanical philosophy, alchemy embraced a radically different view of nature. Where the one denied the reality of life and looked upon animals and plants as complicated machines, the other extended the organic outlook over every facet of nature, including the generation of metals which it believed grew in the earth. As I mentioned, what was probably Newton's earliest independent effort in alchemy was a paper on the vegetation of metals. Especially, alchemy asserted that all things and above all, metals, are generated by the union of male and female. One of Newton's favorite materials for experimentation in the 80's was a substance he called the net, a regulus made from antimony, copper, and iron. In the imagery of alchemy, Venus represented copper and Mars iron. The net referred to the story in ancient myth of a net of gold woven by Vulcan, the husband of Venus, who had good cause to suspect both Mars' intentions and his wife's reception of them, and with the net captured the pair *in flagrante delicto*. One might search some time to find a more suggestive image than the net; it represented the alchemical hermaphrodite, the union of the male and female principles. To the image of sexual generation alchemy frequently joined a second image of purification. Before they could join in effective union, both male and female, animating soul and animated body, had to be purged of the encumbering feces that weighed them down. Alchemists purified their materials eternally; much of the practical

work in the laboratory concerned itself with this task. On occasion,
the juncture of the two themes, generation by male and female and purification, could produce some striking imagery. "I say our true Sperm
flows from a Trinity of Substances in one Essence," Philalethes wrote,
"of which two are extracted out of the Earth of their Nativity by the
third, and then become a pure milky Virgin-like Nature drawn from the
Menstruum of our Sordid Whore." Having seen and duly marked this passage, I found that Newton had discovered it before me and liked it well
enough to repeat it at least seven times in his papers.[37]

Associated with its organic conception of nature alchemy also
thought in terms of active principles, embodied most perfectly in the
philosophic sulfur (inevitably male in that age), which animated the
passive (female) matrix, the philosophic mercury. To the mechanical
philosophy active principles were anathema. Matter in its view was
wholly inert, incapable of initiating any activity or change, subject
to being moved wherever an external impulse directed it. Chemistry almost cried out aloud against the effort to reduce it to such mechanical
terms. Repeatedly its phenomena seemed to reveal foci of activity in
matter. Two cold materials gently mixed suddenly grew hot. Two sub-
stanced joined together to form a compound but shunned a third. It is
striking how readily active verbs inserted themselves into Newton's
descriptions of experiments. An acid solution "wrought upon" spelter
until it dissolved it, and another solution "fell a working wth a sud-
den violent fermentation." Spirits "drew" or "extracted" the salts
from metals. When substances combined, one "laid hold" on the other;
when two sublimed together, one "carried up" the other, and if they did
not sublime, it was because one "held" the other "down."[38] Phenomena
such as these gave great trouble to mechanistic philosophers forced to
explain all of nature with only the categories of size, shape, and motion. In his alchemical reading Newton constantly met the concept of
active principles in nature, and in his laboratory he constantly witnessed phenomena that did seem to require them.

From an early time, the speculations on the nature of things by
Newton, who had begun his scientific career by conversion to the mechanical philosophy, took on a special appearance as he wove the concept of an active-passive dichotomy into them. In the late 60's, his
essay on *The Vegetation of Metals* distinguished nature's two modes of
action, the vegetable and the mechanical, in terms of the active-passive dichotomy. The principles of nature's vegetable actions are

seeds, "her only agents, her fire, her soule, her life." A seed is
never more than a tiny part of the whole surrounded by "dead earth &
insipid water." The grosser substances act as vehicles only, and they
take on different appearances as their particles are moved about.
These changes, Newton argued, are purely mechanical. One can get a
different hue from the mixture of two colored powders, for example, or
one can coagulate milk into butter by agitating it. Although vulgar
chemistry can produce impressive displays, all of her operations involve nothing but the conjunction and separation of particles, that is,
mechanical alterations, and nature employs the same means to produce
the same effects.

> But so far as by vegetation such changes are
> wrought as cannot be done w^thout it wee must
> have recourse to som further cause And this
> difference is vast & fundamental because noth-
> ing could ever yet be made w^thout vegetation
> wch nature useth to produce by it. ...There
> is therefore besides ye sensible changes
> wrought in ye textures of ye grosser matter
> a more subtile secret & noble way of working
> in all vegetation which makes the products
> distinct from all others & ye immediate seat
> of thes operations is not ye whole bulk of
> matter, but rather an exceeding subtile &
> inimaginably small portion of matter diffused
> through the mass wch if it were separated
> there would remain but a dead & inactive earth. [39]

The active-passive dichotomy expressed itself in the "Hypothesis of
Light" in 1675, both by presenting the aether as the active spirit of
nature and by dividing the aether itself into the "maine flegmatic
body" and "other various aethereall Spirits..." [40] Moreover, when Newton began to work toward the *Principia* in the autumn of 1684, his early
essay *De motu* did not embrace the principle of inertia but rather presented the motions of bodies as the results of the interactions of
forces external to bodies with forces internal to them, that is, with
active principles animating lifeless matter which plays a wholly passive role. [41]

     Two and a half years later, when he composed the preface for his
now completed *Principia*, Newton briefly described the content of the
three books and his use of the force of gravity in Book III to explain
celestial motions. "I wish we could derive the rest of the phenomena
of Nature by the same kind of reasoning from mechanical principles,"
he continued, "for I am induced by many reasons to suspect that they
may all depend upon certain forces by which the particles of bodies,

by some cause hitherto unknown, are either mutually impelled towards
one another, and cohere in regular figures, or are repelled and recede
from one another."[42]   At one point he had planned to say much more a-
bout these other forces.   He had drawn up an extended "Conclusio" to
his work, which described at some length the evidence for the interpar-
ticulate forces.   "Hitherto I have explained the System of this visible
world," it began, "as far as concerns the greater motions which can
easily be detected.   There are however innumerable other local motions
which on account of the minuteness of the moving particles cannot be
detected, such as the motions of the particles in hot bodies, in fer-
menting bodies, in putrescent bodies, in growing bodies, in the organs
of sensation and so forth.   If any one shall have the good fortune to
discover all these, I might almost say that he will have laid bare the
whole nature of bodies so far as the mechanical causes of things are
concerned."[43]

        To support this assertion Newton drew primarily upon the evidence
of chemistry.   He was greatly impressed by reactions which produce
heat.   "If spirit of vitriol (which consists of common water and an
acid spirit) be mixed with Sal Alkali or with some suitable metallic
powder, at once commotion and violent ebullition occur.   And a great
heat is often generated in such operations.   That motion and the heat
thence produced argue that there is a vehement rushing together of the
acid particles and the other particles, whether metallic or of Sal Alkali; and the rushing together of the particles with violence could not
happen unless the particles begin to approach one another before they
touch one another.   ...   So also spirit of nitre (which is composed of
water and an acid Spirit) violently unites with salt of tartar; then,
although the spirit by itself can be distilled in a gently heated bath,
nevertheless it cannot be separated from the salt of tartar except by a
vehement fire."   In addition to reactions in which heat appears in formerly cold ingredients, reactions displaying selective affinities also
suggested forces between particles.   When bodies dissolved in acids are
precipitated by salt of tartar, he argued, "the precipitation is probably caused by the stronger attraction by which the salt of tartar
draws those acid spirits from the dissolved bodies to itself.   For if
the spirit does not suffice to retain them both, it will cohere with
that which attracts more strongly."[44]

        Newton did not discover these reactions; they were well known to
the chemists of his day.   All or virtually all of them appeared, for
example, in the writings of Robert Boyle, with which Newton was

familiar. What Boyle fitted into the corpus of his mechanical chemistry became for Newton evidence of active principles by which particles
of matter attract or repel each other. He could not have found this
conclusion in Boyle. Nor could he have found it, in the form in which
he cast it, in alchemical writers. In them he could have found a concept of active principles associated with analogous phenomena, however.
Suffice it to say that without exception the chemical phenomena cited
in his "Conclusio" of 1687 appeared among his alchemical papers of the
previous decade.

In the *Principia* itself, the active-passive dichotomy, recast from
its original form, reappeared in the duality of impressed forces external to bodies and internal *vires inertiae* whereby they resist the effort of external forces to alter their state of rest or motion. In
Query 31 of the *Opticks* Newton explicitly drew out the active-passive
imagery behind his conception.

> The *Vis interiae* is a passive Principle by which
> Bodies persist in their Motion or Rest, receive
> Motion in proportion to the Force impressing it,
> and resist as much as they are resisted. By this
> Principle alone there never could have been any
> Motion in the World. Some other Principle was
> necessary for putting Bodies into Motion; and now
> they are in Motion some other Principle is neces-
> sary for conserving the Motion.

Newton continued with the argument by showing first, from the composi-
tion of motions, that the total quantity of motion is not universally
conserved, and second, that motion is constantly lost through friction
and imperfect elasticity.

> Seeing therefore the variety of Motion which we
> find in the World is always decreasing, there is
> a necessity of conserving and recruiting it by
> active Principles, such as are the cause of
> Gravity, by which Planets and Comets keep their
> Motions in their Orbs, and Bodies acquire great
> Motion in falling; and the cause of Fermentation,
> by which the Heart and Blood of Animals are kept
> in perpetual Motion and Heat; the inward Parts of
> the Earth are constantly warm'd, and in some
> places grow very hot; Bodies burn and shine,
> Mountains take fire, the Caverns of the Earth are
> blown up, and the Sun continues violently hot and
> lucid, and warms all things by his Light. For we
> meet with very little Motion in the World, besides
> what is owing to these active Principles. And if
> it were not for these Principles, the Bodies of the
> Earth, Planets, Comets, Sun, and all things in them,
> would grow cold and freeze, and become inactive
> Masses; and all Putrefaction, Generation, Vegetation

and Life would cease, and the Planets and Comets
would not remain in their Orbs.[45]

While I find the references to putrefaction, generation, vegetation, and life suggestive, there are other passages in which I perceive
a more direct filiation between the Newtonian conception of force and
alchemical ideas. Newton held that the refraction and reflection of
light are caused by forces exerted by refracting and reflecting media
upon corpuscles of light. In the *Principia*, he demonstrated how an attraction normal to a refracting interface entails the sine law of refraction, and in the *Opticks* he drew up a table relating the refractive
indices of various media to their specific gravities.[46] As he noted,
there appeared to be two distinct classes of refracting bodies, those
which abound in "sulphureous oily Particles" and those which do not.
"Whence it seems rational," he concluded, "to attribute the refractive
Power of all Bodies chiefly, if not wholly, to the sulphureous Parts
with which they abound. For it's probable that all Bodies abound more
or less with Sulphurs. And as Light congregated by a Burning-glass
acts most upon sulphureous Bodies, to turn them into Fire and Flame;
so, since all Action is mutual, Sulphurs ought to act most upon
Light."[47] In alchemy, sulphur was the ultimate active agent. It appears to have retained that role in Newton's universe.

De natura acidorum, a paper written in the 90's related sulphur to
another of the active substances of the alchemical world, acids, the
dragons and serpents which devoured sundry kings and queens in numerous
alchemical texts. In this case Newton suggested that the activity of
sulphur, presumably common rather than philosophic sulphur, derives
from the acid it contains. "For whatever doth strongly attract, and is
strongly attracted, may be call'd an Acid." One sentence in *De natura
acidorum* pointed at the role of the alchemical active principle in the
Newtonian conception of force. "The Particles of Acids...," he asserted, "are endued with a great Attractive Force; in which Force their
Activity consists; and thereby also they affect and stimulate the Organ
of Taste, and dissolve such Bodies as they can come at."[48] One offspring of the forces between particles was a radical new conception of
body which reduced it largely to empty space sparsely seasoned with
diaphanous threads of matter. Significantly, he drew upon his alchemical material to describe it by the image of a net.[49]

I trust that my argument will not be misinterpreted. I am not
contending that Newtonian science is a disguised form of alchemy. If

anyone seeks to place that meaning on my paper, I repudiate it with
scorn. Nor am I asserting the more confined argument that Newtonian
attraction is merely an alchemical active principle. Rather I am seeking to find a source that might have stimulated Newton's mind toward
what became at once his break with the prevailing orthodoxy of 17th-
century science and the concept which enabled him to carry that science
to its highest level of achievement. Newton started his career within
the school of mechanical philosophy which rejected any idea of attraction as an occult throwback to a misguided past. Such an idea formed
the very core of his *Principia*. I argue that in alchemy Newton found
a school of natural philosophy which embodied concepts alien to the
mechanical philosophy, and specifically the concept of active principles which animate passive matter. Alchemical active principles are
not Newtonian forces, however. Whatever we do, we must not treat one
of history's rare giants as the passive recipient of others' ideas.
Above all else, Newton himself was an active principle who reshaped
what he received into products which were his own. He exercised his
alchemy on alchemy itself. The idea of attraction does show up in some
alchemical writers, including Sendivogius, whom Newton read carefully-
ly.[50] Nevertheless, the general idea of attraction is not Newtonian
force. It has not yet acquired that exact mathematical definition
which made the concept, not an occult notion, but the heart of a powerful quantitative science. This Newton gave it, transforming the alchemical notion at least as much as the notion itself transformed mechanical philosophy.

There has been considerable research in the last few years on the
influence of the Cambridge Platonists, especially Henry More, on Newton, especially in connection with the concept of active principles.[51]
I do not see that what I propose today conflicts with that research.
There is no reason why a number of sources known to have influenced
Newton could not work in rough harmony toward the same end. Let me
only insist that the evidence of Newton's involvement in alchemy is not
conjectural. It exists in the form of more than a million words in his
hand. In contrast, his notes on More and the Cambridge Platonists do
not stretch even close to a tenth of this amount. Whatever the influence of the Cambridge Platonists, Newton's career in alchemy was real
and extensive and, the papers indicate, reached its peak at the very
time of the *Principia*.

From the point of view of alchemy, the *Principia* was an interruption. Indeed Newton paused in the spring of 1686, when he had not yet put either Book II or Book III into their final forms, to perform a series of alchemical experiments.[52] And once the book was out, he returned to his manuscripts and to his laboratory with undiminished intensity. I find in the continuation of alchemical research in the early 90's powerful confirming evidence that Newton saw the Art in harmony rather than in conflict with his masterpiece. As far as I can judge from the hand, about half of his alchemical papers come from the five year period following the *Principia's* publication.

Among other things, Newton introduced Fatio de Duillier, the young Swiss mathematician who entered intensely into his life at this time, to the Art, and both Fatio and alchemy figured in the mounting emotional tension that wracked Newton's life in 1693. In the summer of that year, sometime after his reception of Fatio's letter of 13 May which he cited, Newton composed the essay *Praxis*, which I have called possibly his most important alchemical paper. At the climax of *Praxis*, Newton claimed to have achieved multiplication.

> Thus you may multiply each stone 4 times & no more
> [he wrote] for they will then become oyles shining
> in y^e dark & fit for magicall uses. You may fer-
> ment them with $\odot$ & $\mathbb{D}$ by keeping the stone & metall
> in fusion together for a day, & then project upon
> metalls. This is the multiplication in quality.
> You may multiply it in quantity by the mercuries
> of wch you made it at first, amalgaming y^e stone
> with y^e $\mathrm{\breve{y}}$ of 3 or more eagles & adding their
> weight of y^e water, & if you designe it for
> metalls you may melt every time three parts of $\odot$
> wth one of y^e stone. Every multiplication will
> encrease its vertue ten times & if you use y^e $\mathrm{\breve{y}}$
> of y^e 2^d or 3^d rotation wthout y^e spirit perhaps
> a thousand times. Thus you may multiply to in-
> finity.[53]

I do not think that we should take this passage seriously, certainly not as evidence that Newton attained the alchemist's goal, but no more as evidence that he believed he had. Newton was overwrought at the time. Within the following three months he wrote the famous letters to Locke and Pepys, which both of them took as evidence of some derangement, as everyone who reads them must. Rather I offer the passage as an indication that twenty-five years of deep involvement in the Art reached their culmination in what was also, for a number of reasons, the tragic climax of Newton's life.

From dated experiments we know that Newton continued to work in
his laboratory after 1693. Although most of the papers are not dated
and cannot be placed with precision, there is no reason to think that
some of them do not also come from the years immediately following the
breakdown. There are at least four scraps related to business at the
Mint during his early tenure there.[54] Sometime between 1701 and 1705
he purchased a number of alchemical works along with other books.[55]
Nevertheless, we cannot avoid the fact that Newton did turn away from
alchemy. If, as I just noted, there are four scraps with Mint business
on them, the fact that there are only four is significant. Only three
of the alchemical books in his library have an imprint after 1700, and
two of those, bound together, were the gift of their author, William
Yworth, to whose support, it appears, Newton did contribute.[56] Of the
major intellectual pursuits of his Cambridge years, alchemy alone failed to follow him to London to participate in the final chapter of his
life. Only recently has the full import of this fact impressed itself
upon me, and I wish to suggest that his turning away from alchemy is a
reality of no less importance than his involvement in it in the first
place. Unfortunately he left no single word, as far as I have been
able to find, to explain his decision. One is forced back entirely upon speculation. Since I have been trying to insist in this paper that
I am speaking from solid evidence and not from tenuous hints, I want to
insist doubly that what is purely speculation be labelled as such. Two
possible reasons, not mutually exclusive, for Newton's final rejection
of alchemy occur to me. First, he may indeed have become disillusioned
with the Art. When he returned to some semblance of normality in the
autumn of 1693, the man who prided himself on separating demonstrations
from mere hypotheses may have been shocked to read what he had claimed
in the *Praxis*. The manic exaltation of the summer had apparently included an alchemical dimension as he seemed finally ready to embrace
the Venus of his dreams, Truth, in all its seductive appeal. In the end
Newton's whole active career in science dissolved away in the following
depression; the alchemical dream may well have dissolved with it. But
I would suggest as well a second, related, possibility. As the full
measure of his achievement in the *Principia* was borne in upon him in
his role as doyen of British science, Newton may have realized that in
fact he had embraced Truth, not quite the enchanting Venus of his alchemical dream, but a very satisfactory mistress nevertheless. He had
extracted the essence of alchemy; the Art itself counseled him to

reject the dross that remained. With the quantified concept of force
he had set science on a new track. Alchemist in one sense to the end,
he projected that concept into the ready soil of natural philosophy.
It has continued to multiply since then beyond what even his wildest
dreams could have imagined.

FOOTNOTES

1 Antonio-Schinella Conti, *Prose e poesi*, 2 vols., (Venice, 1739-56), 2, p. 26.

2 Roger North, *The Lives of the Norths*, ed. Augustus Jessopp, 3 vols., (London, 1890), 3, p. 15.

3 *The Philosophical Works of Descartes*, tr. Elizabeth S. Haldane and G. R. T. Ross, 2 vols., (New York, 1955), 1, p. 191.

4 Galileo, *Dialogue Concerning the Two Chief World Systems*, tr. Stillman Drake, (Berkeley, 1962), p. 462.

5 See their correspondence after 1687, in which the *Principia* figured very prominently, in Christiaan Huygens, *Oeuvres complètes*, 22 vols., (La Haye, 1888-1950), 9 and 10, *passim*.

6 *Oeuvres complètes de Molière*, ed. Gustave Michaut, 11 vols., (Paris, 1947-49), 10, p. 186.

7 *Philosophical Works*, 1, p. 47.

8 *Oeuvres de Descartes*, ed. Charles Adam and Paul Tannery, 10 vols., (Paris, 1897-1913), 8, pp. 144-56, 275-311.

9 Cambridge University Library, *Add. MS. 3970.3*, ff. 473-4.

10 *Add. MS. 3996*, f. 102.

11 *Principles of Philosophy; Oeuvres*, 8, pp. 212-14.

12 *Add. MS. 3996*, f. 121^v.

13 *The Correspondence of Isaac Newton*, ed. H. W. Turnbull, J. F. Scott, A. R. Hall, and Laura Tilling, 7 vols., (Cambridge, 1959-77), 1, pp. 362-85.

14 Fatio to Huygens, 14 June 1687, Huygens to Fatio, 1 July 1687; *Oeuvres*, 9, pp. 168-9, 190-1.

15 *Principia*, tr. Motte-Cajori, (Berkeley, 1934), pp. 164, 192.

16 *Opticks*, (New York, 1952), pp. 350-2.

17 Cotes to Newton, 18 March 1713; *Correspondence*, 5, p. 392.

18 Bodleian Library, *MS. Don. b. 15*.

19 *Add. MS. 3975*, p. 51.

20 Notebook in the Fitzwilliam Museum, Cambridge, n.p. Notes on Basil Valentine, King's College, Cambridge, *Keynes MS.* 64. Notes on Sendivogius, *Keynes MS.* 19. Notes on Philalethes *Keynes MSS.* 51

and 52. Notes on Maier, *Keynes MS.* 29. (*Keynes MS.* 29 contains the material that Newton used in his letter of 18 May 1669 to Francis Aston; *Correspondence*, 1, p. 11.) *MS. Var.* 259 in the Jewish National and University Library, Jerusalem, also contains, along with some other alchemical papers, early notes on Artephius, Flamel, Sendivogius, d'Espagnet, Augurello, Philalethes, Hermes (*Tabula smaragdina*), and several pieces in the *Theatrum chemicum*.

21 *Add. MS.* 3975, pp. 80-3.

22 *Keynes MS.* 32.

23 All of the MSS. of the *Index chemicus* except the drafts between the second and third versions are in *Keynes MS.* 30. The drafts are in a MS. in the Yale Medical Library.

24 John Harrison, *The Library of Isaac Newton*, (Cambridge, 1978), p. 8. By my count the proportion of alchemical works was higher than Harrison's figure of 9.5%.

25 *Keynes MS.* 67. Newton's paragraph is on f. 68^V. Notes and copies from the collection are in *Keynes MS.* 62.

26 *Keynes MSS.* 22, 24, 33, 51, 52, 65. A MS. in the Yale Medical Library. Meheux to Newton, 2 March 1683; *Correspondence* 2, p. 386. Newton's memorandum on the visit in 1696 is printed in *Correspondence*, 4, pp. 196-8; the original is in *Keynes MS.* 26; another version is in the Joseph Halle Schaffner Collection, University of Chicago Library.

27 Dibner Library, Smithsonian Institution, *Burndy MS.* 16. *Keynes MS.* 18. (The text and translation of *The Key* can be found in B.J. T. Dobbs, *The Foundations of Newton's Alchemy*, (Cambridge, 1975), pp. 251-5. All of Mrs. Dobbs' book contributes to its explication.) *Burndy MS.* 10. *Keynes MS.* 28.

28 The sheet, now with *Keynes MS.* 30, belongs with *Keynes MS.* 35, which contains the drafts of chapters.

29 *Keynes MS.* 49.

30 *Keynes MS.* 48.

31 *Keynes MSS.* 21 and 53. Babson College, *Babson MS.* 420.

32 Drafts are found in *Keynes MSS.* 40 and 41, *Babson MS.* 417, and *Burndy MS.* 17.

33 *Keynes MS.* 135.

34 *Add. MS.* 3973, f. 17. *Add. MS.* 3975, p. 149.

35 *Add. MS.* 3973, f. 29.

36 David Brewster, *Memoirs of the Life, Writings, and Discoveries of Sir Isaac Newton*, 2nd ed., 2 vols., (Edinburgh, 1870), 2, pp. 300-2.

37 Eirenaeus Philalethes, "An Exposition upon Sir George Ripley's Preface," in *Ripley Reviv'd*, (London 1678), p. 28. Newton's repeti-
   tion of this passage are found in *Keynes MS.* 30; *Keynes MS.* 34,
   f. 1; *Keynes MS.* 35, sheet 4; *Keynes MS.* 48, ff. 16-16^V; *Keynes
   MS.* 51, f. 1^V; *Babson MS.* 420, p. 8.

38 *Add. MS.* 3973, ff. 13, 21, 42. *Add. MS.* 3975, pp. 104-5, 108-
   9, 281.

39 *Burndy MS.* 16.

40 *Correspondence*, 1, pp. 362-85.

41 John Herivel, *The Background to Newton's Principia*, (Oxford, 1965),
   pp. 257-74; translation, pp. 277-89.

42 *Principia*, p. xviii.

43 A. R. and M. B. Hall, eds., *Unpublished Scientific Papers of Isaac
   Newton*, (Cambridge, 1962), p. 333. Newton later redrafted this
   material for his preface (*ibid.*, pp. 302-8), then finally suppressed it. The two essays constitute first drafts of his later
   Query 31.

44 *Ibid.*, pp. 333-5.

45 *Opticks*, pp. 397-400.

46 *Principia*, pp. 226-8. *Opticks*, p. 272.

47 *Ibid.*, p. 275.

48 *Isaac Newton's Papers & Letters on Natural Philosophy*, ed. I. Ber-
   nard Cohen, (Cambridge, Mass., 1958), pp. 257-8.

49 "Conclusio"; *Unpublished Papers*, p. 341. *Cf.* draft preface, *ibid.*,
   p. 303; a draft from the early 90's for a revision of Book III,
   ibid., p. 317, and *Add. MS.* 3965.6, f. 266^V.

50 See Newton's notes in *Keynes MS.* 19, f. 1.

51 See especially J. E. McGuire, "Force, Active Principles, and Newton's Invisible Realm," *Ambix*, 15 (1968), 154-208.

52 A date in the spring of 1686 appears in his record of experiments,
   apparently marking the beginning of a new series. *Add. MS.* 3975,
   p. 150.

53 *Babson MS.* 420, p. 18^a. This passage occurs in a draft. The final version (p. 17) watered it down a bit without retracting its
   central assertion.

54 *Keynes MS.* 13, f. 1^V bis; *Keynes MS.* 56, f. 1. Public Record
   Office, *Mint Papers* 19.5, ff. 42, 54^V.

55 Harrison, *Library of Newton*, p. 9.

56 *Ibid.*, items 1138, 1302 and 1644. See Yworth to Newton, c.1702;
   Correspondence, 7, p. 441.

S. T. COLERIDGE AND THE HUMAN SCIENCES:
ANTHROPOLOGY, PHRENOLOGY, AND MESMERISM

In 1848 a slender volume appeared under the title, *Hints towards
the formation of a more comprehensive Theory of Life*.[1] Although Coleridge's name was on the title page, the editor, Seth Watson, claimed
in a postscript that the work "might with more propriety be considered
as the joint production of Mr. Coleridge and the late Mr. James Gillman
of Highate." Gillman, a surgeon, had been Coleridge's friend, physician, and admirer for almost twenty years before the latter's death,
and the *Theory of Life* had been written in response to Gillman's wish
to submit an entry for the Jacksonian Prize of the Royal College of
Surgeons in 1816.[2]

Examination of related manuscripts, however, together with the
contents of the *Theory of Life*, make it entirely clear that the work is
Coleridge's in method, structure, style, and sources. It consists of
three tolerably distinct parts, the first logical and semantic, the
second philosophical, and the third zoological and physiological. The
basic problem addressed is "that of the Bearings of Generic Nature on
Germinal Power as the Principle of Individuality."[3] There is a good
deal in this statement of the problem that needs to be unpacked. In
what way is nature generic? What does Coleridge mean by power in general, and by germinal power in particular? What is the principle of
individuality? What context of debate makes such language intelligible,
let alone appropriate? And what, given the title of my paper, has any
of this to do with the human sciences?

Let me try briefly to answer these questions, as a prelude to considering Coleridge's discussion and interrelation of the three sciences
of anthropology, phrenology, and mesmerism. First, Coleridge's view of
nature is directed by his rejection of empiricist philosophy and sensa-
tionalist psychology. Systems in which knowledge was derived through
the senses appeared to him as literally superficial, recognizing only
the surfaces of things, and consequently as sterile and fragmented.
Sciences based on the description of phenomena could not account for
the invisible realm, other than by postulating an implausible, inadequate, and, preposterously, an infinite regress. That, at any rate,
was how Coleridge viewed the principle of transdiction embodied in

Newton's rules of reasoning in philosophy. Where, in such a system, could one find causes that went beyond Hume's constant conjunction? Where, again, were the sources of activity in nature? Newton's atoms were essentially passive, gravitation being superadded to them precisely to avoid placing an active, causal principle at the heart of material nature - in short, to avoid Spinozism. But if activity was not essential to nature, then nature was lifeless, and its disparate atoms made up a dead world of little things.[4]

Besides, Coleridge was a poet and philosopher, for whom imagination and reason - the life of the mind - were all important. He felt an imaginative sympathy with nature, and moreover believed in the possibility of a rational apprehension of nature's laws. He was intellectually and emotionally a Platonist, one who believed that ideas were not merely regulative, but also constitutive. But in the mechanical sciences, ideas were at most regulative, and mind, as Coleridge put it, was merely a lazy looker-on.[5]

He had more than philosophical objections to a system of the world in which mind was passive. He had medical ones too. He knew from his own experience, mercilessly and courageously recorded in his notebooks, of the interaction and somehow the unity of mind and body. The horrors induced by opium, especially during deprivation and withdrawal, were painfully and intimately experienced. The correlation between bodily and mental health was of the greatest interest to him, and underlay his examination of Brunonian physiology, and his concern with Beddoes's writings on the relation between health and education. Psychosomatic phenomena fascinated him, and he had early witnessed, with Davy, a psychosomatic cure in Beddoes's Pneumatic Institution in Bristol.[6]

Body and mind in man were related - were, in as yet undefined ways, *one*. Nature was constituted by ideas, and man was capable of rationality - it followed that man was in some sort a microcosm of nature, and that mind was active and formative in nature as in man. There was, of course, a philosophical scheme that around 1800 gave support to such a view of man, mind, and nature: this was the *Naturphilosophie* and the complementary transcendental idealism of Schelling and his followers. In his transcendental idealism, Schelling started with selfconsciousness and from it deduced nature, arguing that mind produced nature through imagination. *Naturphilosophie* sought to discover the forms of the laws governing nature by an examination of products in nature. These two approaches, from the ideal to the real, and from the real to

the ideal, were brought together by Schelling's doctrine of the ultimate unity of subject and object, mind and nature. Coleridge came to
see Schelling's thought as imperfect, for by making nature absolute it
identified God with nature, and became just another form of Spinozism.
But even when Coleridge had perceived this and other limitations of
Naturphilosophie, he continued to recognize the role it gave to mind,
and to use and develop metaphors and concepts that Schelling had formulated.

Chief among these concepts was the productivity or generic aspect
of nature. Nature arose from the productive tension of mind and its
products. Mind in one aspect appeared as a positive principle, striving forwards, while resisted by the intransigence of its products. One
could even say that this tension and its dynamic reconciliation constituted nature. Polarity was the metaphor that Coleridge, following
Schelling, saw as the essential characteristic of nature, giving rise
to it and ensuring its development. So the answer to the first question, "In what way is nature generic," is, essentially, through its
productivity and through its polar unity and tension with mind.[7]

This metaphor demands dynamic elaboration. *Naturphilosophie* seeks
the form of the laws governing nature. It presents these laws as consequences of the operation of polar powers that are themselves constitutive and causal principles. Coleridge, while modifying Schelling's
account and that of Schelling's disciple Heinrich Steffens,[8] nevertheless adopted the need for a scheme of polar powers. He considered that
if we could identify "the primary constituent *Powers* of Nature," and
could discover "the Forms in which these Powers *appear* or manifest themselves to our Senses," we would have grasped the shape of nature and
its mode of growth. These powers were noumenal, not phenomenal, and
could be regarded as self-actualizing ideas; and things were, "in order
of Thought," subsequent to the powers that produced them.[9] When Coleridge wrote of "Germinal Power," he meant that constitutive, productive,
and actualizing principle in which lies the origins of living things
and their development - he meant, in short, life as a power, a potency
capable of realizing itself.

As I pursue this exegesis, I am constantly aware of the ambition
of the metaphors, of their resistance to any approximation with the
Cartesian sanities of clear and distinct ideas, and of their seeming
irrelevance to any detailed or even general study that could widely
and unequivocally be termed scientific. Before I proceed to deepen the

hole that with Coleridge's aid I have dug for myself, let me remark
that Coleridge's whole intellectual life was a striving after the unrealisable goal of the self-integration of all knowledge, of philosophy
with science, history, theology, and literature; and that his partial
successes are impressive,[10] his failures instructive. He was convinced
that to restrict one's discourse, especially at the outset, within the
realm of clear and distinct ideas, was to ensure that one would never
apprehend new ideas, and that one would remain always at the level of
fancy or understanding - a logical or mechanical faculty - without ever
using reason or imagination, man's creative intellectual faculties. He
was, however, both a lover and student of the fine detail of nature,
and a philosopher determined to subject his metaphysical constructions
to the test of scientific observation. He repeatedly acknowledged that
an hypothesis could be falsified by a single observation. His most
general intellectual constructs were applied to specific problems of
contemporary science; and, while it is true that he, like the cranio-
scopists whose work he examined, sought more for confirmation than for
falsification in the empirical realm, his awareness of the fragility
of hypotheses and of their vulnerability in the face of evidence is un-
usual in the early nineteenth century. He was, as we shall see, often
sceptical where his contemporaries were credulous. He was also careful
to study accurate compendia in chemistry, physiology, zoology, and
other sciences. His notebooks witness to a concern with the precise details of, for example, natural history and comparative anatomy.[11] For
the present, however, I must return to generalities.

Coleridge wrote of the germinal power or the power of life as the
principle of individuality. To write of life as a power was to underline that it "is not a *thing* - a self-subsistent *hypostasis* - but an
act and process." Specifically, it was "the power which unites a given
all into a *whole* that is presupposed by all its parts."[12]

The logical part of the *Theory of Life* is devoted to illustrating
the inadequacy and even the tautological nature of familiar definitions
of life. The philosophical part of the book discusses powers, polarity, productivity, and ascent in nature. The final section, that dealing with zoology and physiology, traces the power of life in its ascent
from the lowest realms to its culmination in man. I am not concerned
here with the logical part of the *Theory*. On a previous occasion in
this university, and elsewhere in print, I have discussed Coleridge's
scheme of the ascent of powers in nature.[13] I hope that the briefest

exegesis of this ascent will suffice now, as a step towards understanding his view of the human sciences.

The basic scheme of the powers is pentadic and hierarchical. The pentad is derived from Coleridge's polar logic, which begins from an initial *prothesis*. This generates the polar opposites of *thesis* and *antithesis*, whose dynamic reconciliation or mid-point Coleridge terms their *indifference* or *mesothesis*. Thesis and antithesis jointly lead to a new *synthesis*, the final stage in the logical pentad, whose parts are related in thought by a productive sequence. But the powers in nature have their origin in the rational activity of mind in its productivity. Coleridge, borrowing heavily from Heinrich Steffens while modifying what he borrows, accordingly constructs a pentad of powers to describe the logical relations of powers in nature. Using Schelling's fundamental metaphors for the construction of matter, Coleridge represents the typical pentad as a cross, in which the vertical or North-South axis corresponds to the polarity of the magnet, the West-East axis corresponds to electric polarity, and the point of intersection of the axes is identified with galvanism or the power of chemical affinity. Schelling had identified magnetism, electricity, and galvanism with length, surface, and depth or inward power respectively, thereby erecting a complex cosmic metaphor. Coleridge translated this into a pentad describing the operations and powers of nature at every level, whose laws were always of the same form, but which were nonetheless distinct in kind. He drew up several variant schemes; an early and typical one ascends through the following stages: ideal, cosmical, geological, potential (i.e. physical), chemical, vital and organic.[14] In the *Theory of Life*, he made it clear that these stages in ascent were related like the rungs of a ladder, which were as clearly separated as animals were from vegetables or men from apes. And yet the identity of the form of the powers at each level conferred a unity on the system, in which the power of life, surging and ascending from level to level, produced ever higher organisms and ever greater individuation, until man, the crown of creation, was reached.[15] It should be emphasized that Coleridge's ascent of life and ranking of species is an intellectual and not an historical one; his classification of organic beings has a basis in a classification of powers and of the degrees of individuation, but has no remotely evolutionary implications.

The preliminary questions that I raised at the outset are now almost answered. The *Theory of Life* has a coda, almost an appendix, on

the powers of magnetism, electricity, and galvanism. But its real conclusion, and its unmistakable culmination, lies in two paragraphs on
man.[16] In the first of these, he discusses the limitations of analogy
when applied to man: "with the single exception of that more than valu-
able, that estimable philanthropist, the dog, and, perhaps, of the horse
and elephant, the analogies to ourselves, which we can discover in the
quadrupeds or quadrumani, are of our vices, our follies, and our imper-
fections." Coleridge's principal reason for postulating the breakdown
of analogy, it turns out, is that he follows Genesis 1:26, in seeing
man as made in God's image, which he interprets as implying self-con-
sciousness, self-government, and the possession of a soul. Analogy, in
short, breaks down in the realms of mind and behaviour.

Anatomically, however, analogy still holds, man being discernibly
the product of a series of converging anatomical tendencies.

"The class of *Vermes*," Coleridge tells us, "deposit a calcareous
stuff, as if it had torn loose from the earth a piece of the gross mass
which it must still drag about with it. In the insect class this resi-
duum has refined itself. In the fishes and amphibia it is driven back
or inward, the organic power begins to be intuitive, and sensibility
appears. In the birds the bones have become hollow; while, with apparent proportional recess, but, in truth, by the excitement of the opposite pole, their exterior presents an actual vegetation. The bones of
the mammalia are filled up, and their coverings have become more simple." Here is a continuous, but not linear, development - taken, it
must be added, fairly directly from a work by Steffens.[17] What follows
also includes sentences from Steffens; but in its purport and conclusions, and in the bulk of its expression, it is pure Coleridge. "Man,"
he continues, "possesses the most perfect osseous structure, the least
and most insignificant covering. The whole force of organic power has
attained an inward and centripetal direction. He has the whole world
in counterpoint to him, but he contains an entire world within himself.
Now, for the first time at the apex of the living pyramid, it is Man
and Nature, but Man himself is a syllepsis," - i.e. a taking together;
a summary (O.E.D.) - "a compendium of Nature - the Microcosm!" I began
by pointing to the identity of mind and nature, the unity of body and
mind, and the unity of powers of mind with powers of nature. Now Coleridge, from a comparative study of skeletons, or rather from Steffens'
account of such studies, has found confirmatory empirical evidence for
these fundamental unities.

"In Man," Coleridge continues, after a brief digression, "in Man the centripetal and individualizing tendency of all Nature is itself concentred and individualized - he is a revelation of Nature! Henceforward, he is referred to himself, delivered up to his own charge; and he who stands the most on himself, and stands the firmest, is the truest, because the most individual, Man. In social and political life this acme is inter-dependence; in moral life it is independence; in intellectual life it is genius. Nor does the form of polarity, which has accompanied the law of individuation up its whole ascent, desert it here. As the height, so the depth. The intensities must be at once opposite and equal. As the liberty, so must be the reverence for law. As the independence, so must be the service and the submission to the Supreme Will! As the ideal genius and the originality, in the same proportion must be the resignation to the real world, the sympathy and the intercommunion with Nature. In the conciliating mid-point, or equator, does the Man live, and only by its equal presence in both its poles can that life be manifested!"[18]

Polarity and dynamism were universal principles for Coleridge, equally applicable in comparative anatomy, physiology, theology, literary criticism, political and social theory, and law. It is worth remembering that his clearest statement of the Law of Polarity occurred in *The Friend* in the midst of a discussion of the opposition between law and religion, where it received chemical illustration.[19] And the meeting place of the realms of mind, matter, and spirit was man, who could properly be studied from a vantage point based in any one of these realms. Man the microcosm was therefore unique in Coleridge's scheme, and his study could not properly be contained within the natural sciences.

There were, however, specifically human sciences that discoursed on man's place in nature, the relation of mind to body, and the nature of the will in its relation to other powers and to body. These sciences brought mental and moral life within the compass of systematic enquiry and empirical exploration. In Coleridge's day, in the early nineteenth century, three of these sciences were especially prominent, generating widespread interest and debate. They were anthropology, phrenology, and mesmerism. In treating them within the context of Coleridge's thought, it is natural to consider anthropology first, as being most closely related to the wider natural sciences, and looking at man as a species proper for the naturalist's scrutiny. Man - man in the

dictionary sense that I have used throughout, embracing woman - man was
part of nature. The whole Romantic movement inclined towards the apprehension of a new nature with and within which man would exist, body and
mind, in active symbiotic and symbolic harmony. As Coleridge remarked
in his *Religious Musings* of 1794:

> Tis the sublime of man,
> Our noontide Majesty, to know ourselves
> Parts and proportions of one wondrous whole!
> This fraternises man, this constitutes
> Our charities and bearings. But 'tis God
> Diffused through all, that doth make all one whole.[20]

This injunction to know ourselves demanded a new scientific enquiry; and
the same springs that led to a recrudescence of cultural rationalism led
also to the interest in anthropology.

Anthropology in Coleridge's day generally had a more restricted focus than its etymology suggests. It was not the whole science of man,
but rather the branch of that science "which investigates the position
of man zoologically, his 'evolution,' and history as a race of animated
beings" (O.E.D.). Contemporary debate concentrated on the problem of
evolution, for many believed that man had evolved into his present
state by degeneration rather than ascent. There was also the need to
clarify ideas about concepts of species, race and variety in their application to man. Coleridge was much interested in these matters, especially in the 1820s, as his manuscripts show.[21]

The issues may conveniently be approached through William Lawrence's article on MAN in Rees's *Cyclopaedia*, 22, 1819. There we find
an examination of the question: *"Does Man constitute a distinct
Species?"* Some writers, notably Monboddo, had asserted that man and
the orangutan belonged to the same species. Lawrence marshalled evidence from comparative anatomy and physiology to show that man is a
species. Following Blumenbach, he defined a species as consisting of
"all animals, which differ in such points only as arise in the natural
course of degeneration." The great variety of mankind arose because
"Man, the inhabitant of every climate and soil, partaking of every kind
of food, and of every variety in mode of life, must be exposed still
more than any animal to the causes of degeneration."

Coleridge made extensive use of Rees's *Cyclopaedia*. He almost certainly read the article on MAN, and may well have surmised if not known
that Lawrence had written this article. This would scarcely have dis-
posed him favourably towards it, for Lawrence was the particular target
of a lengthy polemic in the *Theory of Life*. On the other hand, the

article proposed a pentadic classification of human races, conformable
to Coleridge's dynamic logic, and basing itself on arguments first pro-
pounded by Blumenbach, whose lectures Coleridge had attended at Göttingen in 1799.[22]

Perhaps for these reasons, Coleridge was sympathetic to Lawrence's
arguments here. He accepted the specificity of man, asserting that
"the difference between Man and the larger number of other Mammalia consists in the absence of different species." Within the species, the
principal differences were between races, "as an intermediate term between Species and Variety - the Criterion being that the offspring from
Parents of different Races do in all instances equally, uniformly and
by necessary law inherit & combine the differences of both." The problem then became one of the classification of races and an account of
their origin.[23]

There were various classifications. Cuvier, in his *Règne Animal*,
1817, *1*, 94, identified three especially distinct races, the white or
Caucasian, the yellow or Mongolian, and the negro or Ethiopian. These
were in fact the first three of the five races listed by Blumenbach in
his treatise on the races of mankind, in which he also listed the American and Malay races.[24]

How had these races arisen? Coleridge enumerated different possible answers. First, he found it inconceivable that five distinct
races had been created originally. The notion of ascent from a more
primitive ancestor, Monboddo's "ourang Utang Hypothesis" latterly revived by Oken, he dismissed as "too unphilosophic to require confuta-
tion." James Prichard had advanced a more moderate theory of ascent in
1813, arguing that "the process of Nature in the human species is...the
evolution of white varieties in black races of men." Coleridge, who
knew these arguments, rejected them, favouring instead the view that
the present races of mankind had descended through degeneration from
an original stock.[25] This was perhaps the most commonly accepted view
in Coleridge's day. Kant had proposed four races in 1775. Blumenbach
at first shared his view, but proposed a five-fold division in 1781,
explaining it by degeneration brought about by the influence of climate,
food, and mode of life. Coleridge was happy to accept the pentadic
division of the races, and also the implication that the physical de-
terioration and debilitation of mankind was of moral origin.[26]

A system of five races was amenable to treatment according to Coleridge's pentadic logic, while the invocation of moral causes of

degeneration suggested to him the possibility of correlating the bible
with anthropology. Thus, for example, he opposed the "State of nature,
or the Ouran Outang theology of the origins of the human race" to "the
Book of Genesis, ch. I-X." What was more, he cherished "a hope bordering on a belief, that I can explain the origination and geographical
position of Blumenbach's five Races...on the Mosaic Triad,Shem, Ham,
and Japhet - nay, what is yet more difficult, reconcile the Kantean
diagnostic of *Race* with the Mosaic Documents respecting the Deluge."
Coleridge credited Kant with "the germinal and Substantiating Idea of
Blumenbach's, as it first enabled us to distinguish diversity from variety, or difference in *Kind* from difference in Degree, and to appropriate the term, *Race*, to the former."[27]

Coleridge took Noah's sons as the starting point for the races.
"Still, however, tho' in reverence of my old and dear Friend, *Blumenbach*, I have adopted his...names & mode of applying the theory of *Kant*,
I prefer my own Scheme..." "Yet that...the Book of Genesis and it
alone deserves the name of a History de Originibus Humani Generis, I am
fully persuaded...."[28]

As for the pentad, Coleridge played with Blumenbach's classification of the races, with the Caucasian, supposedly the least degenerate,
as the prothesis, the negro, as thesis, the Malay as the indifference,
the American Indian as antithesis, and the Mongolian as synthesis. "But
in a very short time," Coleridge remarked, "the Indifference was blighted & disfeatured by the divine Judgement - /*Gazed on the Nakedness* of
the Humanity prostrated in the trance of inebriating Nature (Bacchus,
Passion, Lust) - displayed & made merriment & festive worship with the
Lingam [phallus] - /and so became *disguised* - or degenerate. - ". The
races arose because of the moral and spiritual fall of man. Noah and
his sons have so far given us only four races, four fifths of the pentad. "The Synthesis of Shem and Japhet ought to have been the Christian, but [he] exists only *idealiter*. - The other Varieties are in my
scheme the products of Intermarriage, with different properties of
Degeneracy."[29]

Kant's four races are not Blumenbach's five, nor Coleridge's four
actual and one potential. Here is one of the least convincing applications of Coleridge's dynamic logic. But the extent to which the argument is both tentative and paradoxically forced testifies to the unity
of Coleridge's vision.

Anthropology was a part of zoology, at once contributing to and drawing from natural history, comparative anatomy, and physiology. But it was much more. In its bearing upon the history of man and the relations between different races it touched upon religious, political, and social sensibilities. The common assumption that variation derived from moral degeneration not only accorded with Genesis, thus contributing to religious debates about the literal accuracy of the bible and its degree of consonance with science, but also added fuel to controversies about colonialism, imperialism, and slavery, all anathema to Coleridge.

One common means of distinguishing races was by measurements of the skull. Petrus Camper, a Dutch scientist, had proposed a "cephalic index," based upon the inclination of a line drawn from the upper lip to the forehead. "Upon inclining the facial line forward, I obtained a head like that of the ancients; but when I inclined that line backwards, I produced a Negro physiognomy, and definitively the profile of an ape, of a Chinese, or an idiot in proportion as I inclined this same line more or less to the rear." Blumenbach subsequently required additional measurements to be made in order to classify a skull reliably.[30]

We have seen that physical characteristics were associated with moral and intellectual ones. Camper's and Blumenbach's skull measurements were accordingly used to give a first indication of mental and moral endowments. F. J. Gall developed this approach empirically into a system of psychology known as phrenology. For some years he joined forces with J. C. Spurzheim, from whom he separated in 1813. Phrenology was enormously successful in Britain; in 1832 there were approximately thirty British phrenological societies. A large part of the success lay in the connection between phrenology and anthropology. As John Cross wrote in *An attempt to establish physiognomy upon scientific principles*,[31] "if you allow the face of a negro to express less intelligence than the face of a Lord President Blair, then you have granted the truth of physiognomy." Since many of Gall's earlier studies were made in prisons and lunatic asylums, his work had not only racial but also social and penal implications, and even bore on educational issues. Little wonder that phrenology was so widespread and important a science in early 19th-century Britain.

It was based upon five principles: "(1) the brain is the organ of the mind; (2) the mental powers of man can be analysed into a definite number of independent faculties; (3) these faculties are innate, and

each has its seat in a definite region of the surface of the brain;
(4) the size of such region is the measure of the degree to which the
faculty seated in it forms a constituent element in the character of
the individual; (5) the correspondence between the outer surface of the
skull and the contour of the brain-surface beneath is sufficiently
close to enable the observer to recognize the relative sizes of these
several organs by the examination of the outer surface of the head."[32]

The identification of these organs and the divisions of the skull
and brain were empirical and frequently seemingly arbitrary. The evidence for these principles was highly debatable, and was not strengthened by Gall's habit of regarding evidence either as confirmatory of
his views or else as vitiated by other factors. Moreover, Gall's fifth
principle, concerning the correspondence between the shape of the skull
and that of the brain, led to neglect of the study of brain conforma-
tion in favour of the much simpler business of cranioscopy.

These were all limitations of Gall's system, and helped attract
ridicule and abuse. But there were major advantages in his approach.
The scientific revolution of the seventeenth century had achieved the
separation of mind from material nature, made explicit by several writers and enshrined in Cartesian dualism. Then came the rejection of innate ideas, and, through the writings of Locke and Condillac, the construction of a psychology in which knowledge and the faculties themselves were derived from sensations. The divorce of mind from nature,
and its subsequent derivation from the evidence of the senses, were
anathema to Coleridge, who could not accept the sensationalists' epistemological psychology. What was needed was a study of mind pursued
according to the methods of the natural sciences, and recognizing man's
unity with nature.

Gall's approach recognizes these needs, seeking to construct a
biological psychology within the framework of natural history and physiology, and insisting on the unity of man with nature.[33] It was inevitable that Coleridge would take note of the new science of phrenology.
When he first encountered it, however, he was highly sceptical. In a
letter of 1815 to Brabant, a surgeon in Devizes, he wrote that he had
read "Spurzheim's Book," presumably the *Outlines of the Physiognomical
System of Drs. Gall and Spurzheim, indicating the Dispositions and Manifestations of the Mind*, London, 1815. He had found the work "below
criticism." As for Gall's anatomical discoveries, he went on, he could

not judge them, but could show "full half, stated either as Truths or Absurdities," in earlier works.[34]

Spurzheim subsequently visited Coleridge, at Keswick and at High-gate, creating a favourable impression.[35] These visits, together with Spurzheim's co-authorship with Gall, led Coleridge to discuss phrenolo-gy as if it were Spurzheim's invention and science. Meanwhile, the *Edinburgh Review* of 1815 had published a hostile account of the system of Gall and Spurzheim, probably by John Gordon (1786-1818).[36] Gordon developed his objections in *Observations on the Structure of the Brain, Comprising an estimate of the Claims of Drs. Gall and Spurzheim to Dis-covery in the Anatomy of that Organ*, Edinburgh, 1817. Coleridge con-sidered criticism from an Edinburgh Reviewer as a tribute. In August 1817 he observed: "At present I think that Spurzheim is beyond all com-parison the greates Physiognomist that has ever appeared - ."[37] His ac-count of the intellectual faculties had "stood the test of application to an astonishing number of Instances with a most imposing co-incidence - of the moral Indices, I have not the same favorable impression - and his Theory, which is perfectly separable from his Empeiria, I cannot bring into any consistency of meaning....But were it only for the un-doubted splendor and originality of his & Gall's Anatomical Discoveries as to the structure of the Brain...he ought to have been answered, where answerable, with honor & quiet detail of logical objections." His new found sympathy with Spurzheim is also reflected in a footnote in the *Theory of Life*, where he explains that "The Anatomical Demonstra-tions of the Brain, by Dr. Spurzheim, which I have seen, presented to me the most satisfactory proof" of the subtlety of the brain's evolu-tion,[38] i.e. of its structural development.

Coleridge repeatedly insisted that Spurzheim's contribution lay principally in his empirical correlation of mental qualities with the configuration of the skull. Spurzheim was a good naturalist, but his explanations were less persuasive than his observations. The theoreti-cal content of physiognomy was highly problematic. Coleridge's view[39] was reinforced when he read John Abernethy's *Reflections on Gall and Spurzheim's System of Physiognomy and Phrenology*, London, 1821, in which the author stated that nothing but mischief would follow from the accreditation of physiognomy, because it removed responsibility from the individual, now under the determinism of his propensities towards amativeness, destructiveness, and the like.[40] It was all very well for animals, but not for humans endowed with free will.

Such considerations may have made Coleridge progressively more
sceptical as he perceived their force. In 1827 he asserted that "Craniology is worth some consideration, although it is merely in its rudi-
ments and guesses yet."[41] Three years later, the moral and mental implications had come to seem so preposterous as totally to subvert
physiognomical theory: "Spurzheim is a good man, and I like him; but
he is dense, and the most ignorant German I ever knew...when he began
to map out the cranium dogmatically, he fell into infinite absurdities.
You know that every intellectual act, however you may distinguish it by
name in respect of the originating faculties, is truly the act of the
entire man; the notion of distinct material organs, therefore, in the
brain itself, is plainly absurd."[42] One could not look at different
parts of the brain and say, here is a piece of benevolence, and here a
piece of acquisitiveness. Just as Coleridge had rebelled against atomism in the physical sciences, so he rebelled against it in the sciences
of mind. Here again he was going against the majority opinion of his
day, with its faculty psychology. Whether mental faculties were conceived as distinct material manifestations of mind, or as distinct constructs built upon sensation, made little difference; both approaches
fragmented the mind. It is scarcely surprising that Coleridge, as he
perceived this, became increasingly hostile to the system of Gall and
Spurzheim.

In one of his last notebooks, used in 1833-1834, he came out and
attacked phrenology as "anti-scientific - a Quackery, relying for its
apparent Evidence...on a seeming Accumulation of *Coincidence* - ...but
...without in the slightest degree rendering <placing them> in any conceivable relation to each other..." Nevertheless, there was presumably
some relation "causative, or conditional, or instrumental, between the
Brain & the intelligential faculties." Perhaps it might be possible
for phrenology, erected upon a different basis from that of Gall and
Spurzheim, one day to take its place "among the mixed Sciences - i.e.
those Knowledges, which have their root in a universal Truth, tho' they
embody themselves & [derive] their specific forms from Experience..."[43]

Attempts have been made to correlate the teachings of phrenology
with theories of the localization of function in the brain.[44] A good
deal of the evidence for such localization has come from studies of
brain damage. Coleridge too had read about "*cases*, in which certain
parts of the Brain have been wounded, and large portions lost; *yet only
the correspondent faculties* have disappeared, the others remaining."

But those who argued from such cases for the localization of function "had not hit on the *duplicity* of the Organs - and so by a bold assertion of Facts seek to establish the same poly-organism of the Brain, which Spurzheim defends against admitted *contrary* Facts by the assertion of each organ being repeated in each hemisphere."[45] Coleridge might be hostile to the implications of theories of localization of function in the brain. But he was alert and receptive to the evidence, and eager to pursue the relations between mind and brain, mind and nature.

Gall had sought to relate physiology and anatomy to psychology, and this was indeed a part of the unification that Coleridge desired. But he believed that the key to that unification lay in apprehending the unity of powers of mind with powers in nature. Phrenology added little to one's understanding of this unity and its constituent hierarchy of powers. For such understanding, one needed a different science. Mesmerism, with its analogy between the power of the will and that of the magnet, and with its supposed demonstration of the efficacy of magnetism in affecting bodily states, held out hopes for the connectives that Coleridge needed, and appeared as if it might be that vital and different science.

Friedrich Anton Mesmer published the first volume of his *Mesmerismus* in 1814, the year before his death. He believed that he had found a mechanism or influence that pervaded the universe; its uninterrupted flow through the human body was essential to health. "I recognized," he wrote, "that this influence was similar to the one manifested through the properties of the magnet. At the same time, I discovered that the human body is, like iron or magnetic steel, susceptible to similar influences, and that it has poles, that it can act at a distance."[46]

In earlier years, taking the magnetic metaphor literally, he had attempted cures by stroking the bodies of his patients with magnets. After a while, he found that the magnets were redundant, and that he could effect cures by manipulating the magnetic poles of the body without them.

Mesmerism caught on quickly in France. But it was only at the end of Mesmer's life that German physicians, philosophers, and scientists began to take a real interest in his work. The second volume of his *Mesmerismus* appeared in Berlin, 1815, edited by K. C. Wolfart, who published his own account of the subject in Berlin in the same year, when

C. A. F. Kluge also published a report of his investigations into animal magnetism.[47] All these German works were later read and annotated
by Coleridge, whose interest in animal magnetism seems, however, to
date from 1817.

That year saw the publication of the first issue of the *Archiv für
den Thierischen Magnetismus*. In a manuscript dated 8 July 1817, Coleridge raised questions that may well have been inspired by the contents
of the *Archiv*, which we know he had read.[48]

He was never much concerned with Mesmer's theory, but he was anxious that the facts should be fairly considered. "Whence the contemptu-
ous rejection of animal magnetism before and without examination?...the
whole and sole demand is, to examine with common honesty and inward
veracity a series of Facts - and these again not as...mere *historical
Facts*...but the reproducible Facts, facts as strictly analogous to
those of Galvanism as the difference and the continual changes of organic life of the subjects make possible." Stripped of all theoretical
distractions, the sole claim that he perceived the Magnetists as making
was "that the will or (if you prefer it as even less theoric) the *vis
vitae* of Man is not confined in its operations to the Organic Body, in
which it appears to be seated; but under certain previously defined
Conditions of distance and position, and above all of the Patient to
the Agent and of the Agent to the Patient, is capable of acting and
producing certain pre-defined Effects on the living human bodies external to it." Just as the electric eel and other electric fishes could
at will act on bodies at a distance, so, it seemed, could the practitioners of animal magnetism.

Coleridge's invocation of will to explain the phenomena of animal
magnetism underlies his concern with mesmerism. The creation of the
cosmos had been initiated by an act of divine will. The history of
nature could be conceived as a history of mind. The facts of Animal
magnetism offered possible insight into the activity of human will, and
into the relation between powers of mind and powers in matter. Here,
potentially, was a tremendous opportunity for enlarging the sphere of
knowledge. A new power, analogous to galvanism but of a higher order
and subject to the dictates of the will, would render Coleridge's dynamic scheme of the sciences at once more plausible and more noble.
All that he asked in 1817 was the suspension of prejudiced disbelief,
and a fair trial for the magnetists' claims. He intended to try animal
magnetism himself; meanwhile, he would be neutral.

The Friend of 1818 contained a condemnation of animal magnetism;
but in the margin of several copies Coleridge recanted, and stated his
belief that the nervous system of one body could act physically on that
of another.[49] He had in fact already been persuaded of the truth of
the facts of animal magnetism, now admitted by some of its former "illustrious Antagonists."[50]

By 1818-19, when Coleridge came to give his philosophical lectures,
he was fully convinced that the facts of animal magnetism were undeni-
able, attested by the most scrupulous and formerly sceptical observers.
One person could act upon another through will. Here was a possible
explanation of magically induced sleeps and oracles. "I think it...
most highly probable," Coleridge informed his audience, "[that mesmer-
ism] was known to the priests in Egypt - that it was conveyed by tradition to the latest period of the Greek Empire."[51]

The effectiveness of oracles and ancient priests had often been
dismissed as arising from imagination. The success of mesmerism now
testified to the reality of the power of will; if this was attributed
to imagination, then imagination was truly a power to be reckoned with.
"...if the zoö-magnetic influx be only the influence of the Imagination," Coleridge observed in 1819, "the active Imagination may be a
form of the Zoö-magnetic Influence."[52] The association of the new power with will and imagination, determining the fate of the soul and constituting the foundations of creative activity, elevated it to a new
dignity. Henceforth Coleridge usually referred to it as zoomagnetism,
the higher magnetic power in life, and no mere animal magnetism.

He invoked the argument from analogy in support of the claims of
zoomagnetism. There was, for example, "the evident existence of a magical Energy throughout the Creation..., that ...is the ground and pre-
condition of the Mechanical." This energy was most apparent in the organic realm, which revealed "a Power higher than the Organic Function."

Next, if one supposed zoomagnetism to be true, one could offer rational explanations otherwise lacking for a whole series of phenomena,
previously "most unsatisfactorily explained away with Lies, Tricks, or
the Devil, the Oracles of the ancients, Charms, Amulets, witchcraft,
Prophecies, Divination, and extra ordinary...Individuals." Prophecy,
faith healing, and a host of other phenomena could now be confirmed and
explained.

Finally, zoomagnetism seemed to deal a mortal blow "to Materialism, Sensuality, and *Worldliness* on the one hand, and to Superstition

and Credulity on the other." This argued for the truth of zoomagnetism, for "there must be Truth in that, which every good and wise man
must wish to be true: nor need the strongest mind be ashamed of an
argument which beyond all others weighed with Socrates, Plato, and
Tully, in the belief of Immortality." Even the miracles of the Christian faith were confirmed by zoomagnetism, which acquainted us "with
the summits of human nature, on which the Divine, as it were, alighted."

All this constituted a powerful and seductive argument for the
truth of zoomagnetism. Coleridge wanted it to be true. But alas, he
remained doubtful, since the reported facts did not sufficiently bear
out the theory. Even the most disciplined magnetists were guilty of
corrupting facts by their theories. Coleridge deplored this, consistently advocating the separation of fact from speculation. What he now
advocated was essentially an induction - "from the individual Cases,
selecting chronologically such a number as that all the remarkable Phenomena of Magnetized Patients should be found in some one or other of
the Cases selected. - and then drawing from each part the fair inference, of which it is capable, or rather, simply generalizing their import." Throughout Coleridge's scheme of things, the facts were a
touch-stone; theory, in metaphysics, in the philosophy of nature, and
in the natural sciences, had to yield before contrary evidence.

Coleridge remained fascinated by mesmerism for the rest of his
life. It seemed to promise insight into "Life, Mind, Will, Individuality." Yet it constantly failed to provide the longed-for enlightenment.
If only it could be proved true. What then would follow? Nothing less,
Coleridge believed, than the demonstration of the existence of a soul,
capable of acting independently of the bodily organs.

Anthropology, phrenology, and mesmerism, three virtually new disciplines, offered a scientific approach to man that, in supplementing
the more traditional biological sciences, might turn the ancient doctrine of man as microcosm from metaphor into scientific truth. In the
light of Coleridge's intellectual aspirations, his rigorous and critical use of empirical evidence and his insistence on logic and coherence
in argument must have been constantly frustrating. Romantics could be
hard headed and clear sighted - Coleridge was. His metaphysics gave
structure to his thought, without vitiating his grasp of evidence in
the natural and the human sciences.

FOOTNOTES

1 London, 1848; reprint by Gregg International Publishers Ltd., West-mead, Farnborough, Hants., England, 1970.

2 H. J. Jackson, introduction to the *Theory of Life*, to be published in Bollingen Series LXXV, *The Collected Works of Samuel Taylor Coleridge*, general ed. K. Coburn, Princeton University Press and Routledge & Kegan Paul; H. J. Jackson, "Coleridge on the King's Evil," *Studies in Romanticism*, 16, 1977, 337-348; see also chap. 2 of my forthcoming monograph, *Samuel Taylor Coleridge and early 19th-century science*.

3 Coleridge MS Notebook 59 f. 5 [1827].

4 Levere, "Coleridge, Chemistry, and the Philosophy of Nature," *Studies in Romanticism*, 16, 1977, 349-379; *Coleridge and...Science*, chap. 3.

5 Levere, "S. T. Coleridge: A Poet's View of Science," *Annals of Science*, 35, 1978, 33-44. Coleridge to T. Poole, [23 March 1801], *Collected Letters of Samuel Taylor Coleridge*, ed. E. L. Griggs, vol. 2, Oxford, 1966, p. 709. Levere, *Coleridge and...Science*, chap. 4.

6 R. Guest-Gornall, "Samuel Taylor Coleridge and the Doctors," *Medical History*, 17, 1973, 327-342; K. Coburn, "Coleridge: A Bridge between Science and Poetry," in *Coleridge's Variety: Bicentenary Studies*, ed. J. Beer, Macmillan, 1974, pp. 81-100; Bollingen Series L, *The Notebooks of Samuel Taylor Coleridge*, ed. K. Coburn, vol. 1, Pantheon Books, New York and Toronto, 1957, 388 & n; Levere, "Dr. Thomas Beddoes and the Establishment of his Pneumatic Institution," *Notes and Records of the Royal Society of London*, 32, 1977, 41-49.

7 B. Gower, "Speculation in Physics: The History and Practice of *Naturphilosophie*," *Studies in the History and Philosophy of Science*, 3, 1973, 301-356; H. Zeltner, "Das Identitätssystem," in H. M. Baumgartner, ed. *Schelling.Einführung in seine Philosophie*, Karl Alber, 1975, pp. 75-94; T. McFarland, *Coleridge and the Pantheist Tradition*, Oxford, 1969; Levere, *Coleridge and...Science*, chap. 3.

8 Levere, *Studies in Romanticism*, 16, 1977, 360 n41, lists works of Steffens annotated by Coleridge.

9 Coleridge, British Museum MS Egerton 2801 ff. 143-144; marginal note in R. Hooker, *Of the Lawes of Ecclesiastical Politie*, London, 1682, p. 27 (in British Museum).

10 E.g. the three MS vols. of his *Opus Maximum* in Victoria College, Toronto.

11 See especially MS notebooks 27 & 28; the bulk of the scientific entries will be published in vol. 4 of the *Notebooks*, and are discussed in chaps. 5-7 of my *Coleridge and...Science*.

12 *Theory of Life*, pp. 94, 42.

13 "Coleridge and Romantic Science," in *Science, Technology, and Culture in Historical Perspective*, ed. L. A. Knafla, M. S. Staum, & T. H. E. Travers, University of Calgary Studies in History No. 1, 1976, pp. 81-104; *Studies in Romanticism*, 16, 1977, 368ff.; *Coleridge and...Science*.

14 *Notebooks*, 3, 4226; Victoria College, Toronto, Coleridge MS BT 37.

15 *Theory of Life*, pp. 70-85.

16 *Ibid.*, pp. 84-85.

17 H. Steffens, *Beyträge zur innern Naturgeschichte der Erde*, Freyberg, 1801, pp. 315-316; the copy annotated by Coleridge is in the British Museum.

18 *Theory of Life*, p. 86.

19 *The Friend (Collected Works, 4)*, ed. B. Rooke, 2 vols., Princeton, 1969, 1, 948n.

20 *Poetical Works*, ed. E. H. Coleridge, O. U. P., 1912, reprinted 1973, pp. 112-113.

21 See below, & J. H. Haeger, "Coleridge's Speculations on Race," *Studies in Romanticism*, 13, 1974, 333-357.

22 *Theory of Life*, pp. 17 ff., enters into the Lawrence-Abernethy debate on the latter's side. The debate is described in O. Temkin, "Basic Science, Medicine, and the Romantic Era," *Bulletin of the History of Medicine*, 37, 1963, 97-129. Blumenbach's association of the formation of races with degeneration appears in successive editions of his *Handbuch der Naturgeschichte*.

23 Quoted in Haeger, *op. cit.*, pp. 335, 337.

24 Blumenbach, *De Generis Humani Varietate Nativa*, Göttingen, 1775, has four races; the 1781 ed. lists five, as does the German translation, *Über die naturlichen Verschiedenheiten in Menschengeschlechte*, Leipzig, 1793, which Coleridge read. A useful collection is *The anthropological treatises of Johann Friedrich Blumenbach... and the Inaugural dissertation of John Hunter, M. D. on the varieties of man*, trans. and ed. by T. Bendysche, London, 1865.

25 BM MS Egerton 2800 ff. 100^V-101^r. N26 f. 99^V. J. C. Prichard, *Researches on the Physical History of Man*, 1813, ed. and intro. G. W. Stocking, Chicago, 1973, p. liv.

26 I. Kant, *Von den verschiedenen Racen der Menschen*, 1775, reprinted in *Vermischte Schriften*, 1799, 607-632, read and annotated by Coleridge. Coleridge, in *Table Talk*, 24 Feb. 1827, says that Kant proposed 3 races, but elsewhere (marginal note, Kant, *Vermischte Schriften*, 3, 121) correctly mentions Kant's four races. MS Notebook 36 ff. 59^V-60, cf. Haeger, *op. cit.*, p. 338.

27 Coleridge, *On the Constitution of Church and State* (*Collected Works,
    10*), ed. J. Colmer, Princeton, 1976, p. 66. MS Notebook 27 ff.
    80-79^V.

28 Notebook 26 ff. 99^V-100; Notebook 42 f.33.

29 Notebook 26 ff. 99^V, 100. The Excursus Notes to my *Coleridge and...
    Science* will give further information about Coleridge's readings in
    anthropology and the other human sciences.

30 P. Camper, *Dissertation sur les variétés naturelles qui caracteri-
    sent la physionomie des divers climats et des différents ages...*,
    Paris, 1792, pp. 12, 40, cited by Haeger, *op. cit.*, 339, Anthropo-
    logical Treatises, pp. 235-237; Haeger, 339.

31 Glasgow, 1817, p. 13.

32 *Encyclopaedia Britannica*, 11th ed., art. PHRENOLOGY. See also R. M.
    Young, *Mind, Brain, and Adaptation in the Nineteenth Century...*,
    Oxford, 1970. Other valuable literature is discussed in R. J.
    Cooter, "Phrenology: The Provocation of Progress," *History of Sci-
    ence*, 14, 1976, 211-234.

33 R. M. Young, *op. cit.*, pp. 2, 14-16.

34 *Collected Letters*, 4, 613.

35 *Notebooks*, 3, 4269n & 4355n.

36 *Edinburgh Review*, 25, 1815, 227-268.

37 *Notebooks*, 3, 4355.

38 *Theory of Life*, p. 73.

39 *Friend*, 1, 415n.

40 *Reflections*, p. 7.

41 *Table Talk*, 24 June 1827.

42 *Ibid.*, 2 July 1830.

43 Notebook Q, ff. 22, 23, 22^V.

44 E.g. R. M. Young, *op. cit.*

45 British Museum MS Egerton 2801 f.141.

46 *Mesmerismus...*, Berlin, 1815, 1, xvii. See also R. Darnton, *Mesmer-
    ism and the End of the Enlightenment in France*, New York, 1970.

47 Wolfart, *Erläuterungen zur Mesmerismus*, Berlin, 1815; C. A. F.
    Kluge, *Versuch eirer Darstellung des animalischen Magnetismus als
    Heilmittel*, Berlin, 1815.

48 British Museum Add. MS 36532, ff. 7-13, partly printed in *Inquiring
    Spirit*, ed. K. Coburn, Routledge & Kegan Paul, London, 1951, p. 30.

49 *Friend*, 1, 598n.

50 These included Blumenbach and Cuvier.

51 *The Philosophical Lectures of Samuel Taylor Coleridge*, ed. K. Co-
    burn, The Pilot Press, London, 1949, pp. 424, 305.

52 Notebook 29 f. 62.

EVOLUTION, ETHNOLOGY, AND POETIC FANCY:
SIR DANIEL WILSON AND MID-VICTORIAN SCIENCE

In September 1859 Albert, Prince Consort, delivered a speech to a
distinguished body of British gentlemen in Aberdeen, Scotland. Itself,
this was scarcely a remarkable event. Albert gave many speeches during
his lifetime, and many Victorian gentlemen listened patiently to such
addresses. Yet this occasion was somewhat singular. The gathering to
which Albert spoke comprised the elite of the British scientific community, the British Association for the Advancement of Science. Albert
was its president for that year. A friend of British science throughout his life in England, and one of the architects of the Great Exhibi-
tion of 1851, he began his address with a general discussion of the
nature of science - a review which has been described as one "that summarized prevailing philosophical views and did not break new ground."[1]
It was, all in all, a concise reflection of the prevailing consensus
about the nature of mid-Victorian science.

Early the next year, the Prince Consort's presidential address
found its way into the *Canadian Journal*, the main publication of the
Central Canadian scientific community. Those who read Albert's remarks
were treated first to some grand generalizations. "To me," they read,
"Science, in its most general and comprehensive acceptation, means the
knowledge of what I know, the consciousness of human knowledge. Hence,
to know, is the object of all Science; and all special knowledge, if
brought to our consciousness in its separate distinctiveness from, and
yet in its recognised relation to the totality of our knowledge, is
scientific knowledge." The primary activities of the scientist, Albert
continued, are those of analysis and synthesis, the reconstruction of
"a unity in our consciousness," a new understanding of that aspect of
the natural world under observation. "The labours of the man of Science are therefore at once the most humble and the loftiest which man
can undertake," for the process of organized and purposive observation
of phenomena which relates the observed object to "the general universe
of knowledge" - through arrangement and classification - has as its end
nothing less than the discovery of "the internal connexion which the
Almighty has implanted" in previously incongruous elements of nature.[2]

Hence, the central concern of the different sciences is to establish the essential unity of all nature. But this, Albert warned, is an
onerous task because it is also an endless one: "for God's world is
infinite." Onerous or not, he added, it is the duty of the scientist
to remain conscious of the "unity which must pervade the whole of science," and the major way this can be done is "in the combination of
men of science representing all the specialities, and working together
for the common object of preserving that unity and presiding over that
general direction." The true science, the best science, he concluded,
must "proceed...by the inductive process, taking nothing on trust,
nothing for granted, but reasoning upwards from the meanest fact established, and making every step sure before going one beyond it....This
road has been shown to us by the great Bacon, and who can contemplate
the prospects which it opens, without almost falling into a trance similar to that in which he allowed his imagination to wander over future
ages of discovery!"[3]

Albert's declaration points to the important fact that on the eve
of the Darwinian revolution, in English-speaking countries the scientific enterprise was still conducted along lines laid down by Bacon.[4]
This was certainly the case in British North America, where "Bacon,"
like "Reid" and "Paley," was a surname that bore the stamp of final
authority. Baconian science, with the Scottish "Common Sense" philosophy and a Paleyite natural theology, completed the triumvirate for intellectual orthodoxy that dominated many educated Anglo-Canadian minds
for the first three-quarters of the nineteenth century.[5]

The Baconian method was rationalistic and empirical, deduction
after observation, but it did not exclude the Almighty. "And all depends," he had written in the introduction to *The Great Instauration*,
his blueprint for the new science, "...on keeping the eye steadily fixed upon the facts of nature and so receiving their images simply as
they are. For God forbid that we should give out a dream of our own
imagination for a pattern of the world; rather may it be graciously
granted to us to write an apocalypse or true vision of the footsteps of
the Creator imprinted on his creatures."[6] Albert's recapitulation of
the commonplaces of mid-Victorian British Baconian science had not
overtly equated the cause of science with that of religion, except,
perhaps, for its almost gratuitous mention of the Almighty. But there
were many individuals, both in Albert's learned audience and among his
later readers, for whom the Prince Consort's articulation of the

assumptions of that science was little less than royal approval of just
such an equation.

I

Not a few such readers were to be found at the time among those who
practised science in British North America. While men of religion in
Canada sought to lay the appropriate religious foundations for the study of nature through the teaching of their own adaptations of natural
theology, some of the leading scientists in the colony sought, in their
lectures and writings, to build, upon the edifice so carefully constructed by their colleagues, a science in accordance with the orthodox
Christianity of their day. "Let us remember this at least," concluded
the president of the Canadian Institute, Daniel Wilson, early in January of 1860, "that that great ocean of truth does lie before us, and
even those pebbles which our puerile labours gather on its shore, may
include here and there a gem of purest ray; and meanwhile the search
for truth...will bring to each one of us his own exceeding great re-
ward."[7]

Daniel Wilson was, by 1860, one of Canada's leading scientists and
educators. A native of Edinburgh, the forty-four year old Scot had only been in British North America for a half-dozen years. But by then
he had already acquired a respectable international reputation in the
field of science. Wilson had come to science after first considering
careers in art (he studied for a time with William Turner and had considerable talent as an artist) and in popular literature (he had written reviews and essays for magazines such as *Chambers' Journal*, and had
published anonymously a *History of Oliver Cromwell* and a *History of the
Puritans* with the firm of his lifelong friend Thomas Nelson).[8] By the
early 1840s, however, his mind had turned to antiquarian and ethnologi-
cal concerns. In 1847, while secretary of the Scottish Antiquarian
Society, he published the massive *Memorials of Edinburgh in the Olden
Time*, with illustrations executed by himself. Four years later appeared *The Archaeology and Prehistoric Annals of Scotland*, which earned for
its author an honourary Doctorate of Laws from St. Andrew's University
and gave to the world the word "prehistory." The impressively research-
ed volumes which made up this latter work also marked a substantial attempt by the author to effect, in the words of the preface to the first
edition, "the transition from profitless dilettantism to the

intelligent spirit of scientific investigation."[9]   In 1853 Wilson immi-
grated to Canada with his wife to accept a position at the University
of Toronto as professor of history and English literature. By 1859 he
had established himself in the newly built University College, was an
active member of the Canadian Institute, and had become editor of the
Canadian Journal.

The peroration of Wilson's 1860 presidential address to the Canadian Institute had closely paraphrased a famous example of Sir Isaac
Newton's sense of humility in the face of the seeming boundlessness of
nature and the unity of its laws. But as put forward by the president
of the Canadian Institute, that truth and unity were given their essential meaning by the fact that they were, and could only be, a Christian
truth and a Christian unity. "The experience of the past shows how frequently men have contended for their own blundering interpretations,"
Wilson concluded, "while all the while believing themselves the cham-
pions and the martyrs of truth." Such self-delusion would not occur if
one was careful to remember the religious nature of truth, for "all
truth is of God, alike in relation to the natural and the moral law,
and of the former, as truly as of the latter may we say: 'if this counsel or this work be of men, it will come to nought; but if it be of God,
ye cannot overthrow it; lest haply ye be found even to fight against
God.'"[10] A year later, Professor Wilson, still president of the Canadian Institute, reiterated the same theme. The aim of the institute,
he said, should be to investigate the laws of nature and to uncover
"new truths in every department of human knowledge."

A sonnet from a volume of Daniel Wilson's poems published in the
mid-1870s strikes to the heart of this theme:

> I stood upon the world's thronged thoroughfare,
> And saw her crowds pass by in eager chase
> Of bubbles glistening in the morning rays;
> While overhead, methought God's angels were
> With golden crowns of which all unaware
> They heedless crowded on in folly's race.
> But yet methought a few were given grace,
> With heavenward gaze, to aspire for treasures there,
> All trustfully as an expectant heir;
> Through whomme [sic] the soul shone, as the body were
> But a veil, wherein it did abide,
> Waiting till God's own hand shall it uncover.
> O God! that such a prize in vain should hover
> O'er souls in nature to Thyself allied![11]

At one level Wilson's sonnet expresses the common Christian desire to
transcend his finiteness through the pursuit of the eternal; at another

it is a gesture of thanks for the existence of a scientific elect,
those "few [who] were given grace,/With heavenward gaze, to aspire for
treasures there,/All trustfully as an expectant heir." Science thus
could serve as a means of grace for the select few to whom were given
both the ability and the duty to lift the veil with which God had ob-
scured the presence of His hand in nature. This view also finds close
parallels in Wilson's scientific work. To the individual, Wilson wrote
in *Prehistoric Man* (1863), "'The drift of the Maker is dark, an Isis
hid by the veil!'"; but with the aid of the inspiration and guidance of
religion, such darkness is dispelled, for "Christianity...lifts for us
the veil of Isis, tells of the Righter of all the wrongs of ages."[12]

The dictates of Christian piety were not, then, confined only to
the pew on Sundays, not just to the notebooks of philosophy students,
nor to lectures on natural theology. They also found their place in
the field trips taken by professors of science, and they took their
hold as well upon the minds of anyone, academic or layman alike, who
looked to science as a source of religious inspiration and affirmation.
The inductive, Baconian science extolled by Prince Albert required no
abstract "hypotheses" unprovable by simple observation and induction,
no ingenious theories. The Canadian natural theologian, James Bovell,
in fact, had warned his students of the dangers of such unwarranted ex-
tensions of the scientific imagination. "While false systems of philosophy may tantalize and fret the mind," he said, "the calm and reflecting reasoner on revealed truth is content to curb his imagination, and
to accept the creator as He has thought fit to shew himself."[13] The
main requirement was simply a "child-like heart" and the humility that
must necessarily accompany the thought of any Christian who, aware of
the limitations of his own reason, nevertheless yearns for a measure of
unity with the Divine.

Those of such a mind who read Prince Albert's address to the British Association for the Advancement of Science might well have placed
such an interpretation on his remark that the man of science "only does
what every little child does from its first awakening into life, and
must do every moment of its existence; and yet he aims at the gradual
approximation to divine truth itself." He might well have concluded
that by the simple inductive process of observation in the Baconian
tradition, "we thus gain a roadway, a ladder by which even a child may
almost without knowing it, ascend to the summit of truth."[14] Such lessons were not lost on Canadian students only a few years removed from

childhood. Writing in the first volume of the *Dalhousie Gazette*, the
first Canadian university newspaper, one anonymous student declared
that "the great object of science is the ascertainment of causes - the
search after the ultimate, - the making of generalizations - the resolution of plurality into unity." Its "indirect, (although equally
great) object," he concluded, "is the cultivation of all those qualities which dignify and adorn the moral nature of man." Science could
therefore come to the aid of the nineteenth-century moral philosopher
and his insistence on the reality of the "moral nature." It should
therefore come as no surprise to find editorials in Canadian student
newspapers such as the one that wondered: "What can have a greater influence in purifying and elevating the mind of man than the study of
the beautiful and yet wonderful works of Nature around him! What could
raise our thoughts more towards the Supreme Being than the contempla-
tion of Nature, and a knowledge of its wonderful laws."[15]

But such a conception of the methods and aims of science was possible only as long as the essentially static, mechanistic universe of
Newton's physics and Paley's physico-theology remained intact. From
the late eighteenth century on, however, the idea of the immutability
of nature's design had been constantly undermined: by Kant, Buffon,
Laplace, Herschel, Cuvier, Lyell, and others. Work in astronomy, geol-
ogy, and paleontology, in particular, pointed to the fact that nature
itself was mutable, changing. By 1859 the *fact* of organic change was
well known; there was lacking only a theory of the *mechanism* by which
such change took place. "In this development," John C. Greene has observed, "Darwin played a last-minute but decisive role."[16] It was Darwin, more than anyone else in the nineteenth century, who helped to
lift the corner of the veil of Isis. And as with Bluebeard's last
wife, impelled to unlock the last door in the castle, many were aghast
at their first glimpse of what hitherto had been shrouded in darkness.

Daniel Wilson's 1860 presidential address to the Canadian Institute, published in the *Canadian Journal* early that year, was largely a
review of Darwin's *Origin*. For Wilson, science was largely Baconian
in method, but it was also one that marks the hazy frontier which links
science to the religious and poetic sensibility. The science of William Dawson at McGill clearly bore the Hebraic legacy of his Calvinism
(as may be observed by reading Dawson on biblical cosmology); that of
Wilson was equally informed and shaped by Wilson the artist and poet.
"In some respects, and perhaps with truth," the journalist G. Mercer

Adam observed near the end of Wilson's life, "it may be said that Dr. Wilson would have done more justice to himself if he had made a choice in his life's work between literature and science rather than, as he has done, given the prose side of his mind to archaeological studies and reserved its poetical side for literature."[17] There is a significant measure of truth in this statement. But it is rather too artificial and clear-cut. Just as there was no absolutely rigid "scientific" or "religious" side, nor two equal commitments to the different sources of authority implied by these words, there was likewise not, in Wilson's mind any clearly defined "prose" or "poetic" side.

In fact Wilson's poetic sensibility was always a present and determining factor in his thought, including his science. One of his students at University College in the early 1870s took down the following lecture notes:

> Poetry precedes prose...As *Rhetoric* has its own forms etc. distinct from Logic, so poetry has its own....So like pulpit-preaching or pleading - *Poetry aims at giving pleasure by exciting and influencing the Emotions* - and the *license* allowed is only limited by the object the poet has in view....Poetry addresses the *taste* or aesthetic faculty.

Or again:

> Rhetoric is the art of speaking (and writing) well and persuasively the language of Emotion....It employs an elaborateness of structure inadmissible in Logic....
> In Rhetoric - There must be *continuous emotion*, springing from the subject itself, and *true feeling*.[18]

The aesthetic bent of Wilson's mind must constantly be recalled when considering his scientific as well as his literary work; so should the function of rhetoric, thus defined, when assessing the scientific writings of Wilson (and others) in the nineteenth century.

When Wilson spoke of rhetoric, it is obvious that he did not exclude writing upon scientific subjects from its scope. Scientific writing of the Victorian period was very much an art of persuasion. Parts of Wilson's review of Darwin's *Origin* could have served as blackboard examples of that art. Canadian science, he admitted, had necessarily to be in large measure a practical one. Yet any science of true worth must also transcend purely utilitarian concerns. "In the Canadian Institute," he therefore concluded, "it may be presumed that we pursue science for the discovery of its secret truths; that we climb the steps of knowledge, as the traveller ascends the mountain's unexplored cliffs, gladdened at every pause in his ascent with new grandeur and beauty in the widening horizon which opens on his delighted gaze."[19] Here is the rhetoric of romantic aesthetics under the guise of the aims of science, in Wilson's view, science was of value insofar as it excited the aesthetic faculty. Wilson's science in short, owed much to the tradition of which Edmund Burke's essay *The Sublime and the Beautiful* was an integral part. The secret of the veils of Isis as with all veils, lay in the mystery maintained, not in the actual object hidden.

II

Wilson's reaction to Darwin's theory of the origin of species was one characteristic of the day, especially on two of the basic issues raised in the *Origin*: first, that there is in nature no absolute distinction between species and varieties because all organic creation is descended from a common primordial stock; second, that the survival or extinction of species rests primarily upon a process, which Darwin chose to call "natural selection," that led to the "survival of the fittest" (a term which Darwin borrowed from Thomas Malthus and lived to regret).[20] His commitment to an inductive, Baconian science based upon observation meant the rejection, as a consequence, of the notion of a science which set forth "hypotheses" unproven by such observation and resting mainly upon an evidential basis of statistical probability extrapolated from a mass of evidence.[21]

Wilson's scientific judgements were also shaped largely by his essentially moral outlook. As much a man of religion as he was of science (Wilson was a broad church Anglican), his scientific works were profoundly affected by the demands of his religious conscience. And in this mixture of religious and scientific assumptions can be seen the problem of determining the ultimate sources of conviction. Daniel Wilson was drawn to science because for him it held the means by which he hoped to penetrate the veils behind which was what Augustine had called the "great and hidden good."

His basic objection to Darwin was simply that Darwin had put forward conclusions not supported (though not contradicted) by his evidence, substantial though that may have been. Furthermore, like William Dawson, he believed that these conclusions went beyond the scope of legitimate scientific inquiry. "Nothing is more humbling to the scientific enquirer," Dawson began his review of Darwin's *Origin*, "than

to find that he has arrived in the progress of his investigations at a
point beyond which inductive science fails to carry him." Science, he
went on, was an endeavour that rested upon an insoluble contradiction.
"True science is always humble, for it knows itself to be surrounded by
mysteries - mysteries which only widen as the sphere of its knowledge
extends. Yet it is the ambition of science to solve mysteries, to add
one domain after another to its conquests, though certain to find new
and greater difficulties beyond."[22] Paradoxical as it may now seem,
Dawson's science satisfied, rather than frustrated, him. On the one
hand, an inductive Baconian science helped to establish authoritatively
the "truths" of the natural world; yet, on the other, the very boundlessness and seamlessness of the web of nature meant the perpetual revelation of more mysteries of life. Since these were those of God, the
endless search for their solution became, for Dawson, a kind of scientific counterpart of the Christian's quest - even in the knowledge that
his fallen nature makes the mission impossible except through salvation
- for unity with the mind of God, for an insight, however limited and
inadequate, into the Divine will. It was for him the ultimate form of
piety.

From this point of view, the benefit of Baconian empiricism was
that since it was based upon a finite number of observations, limiting
itself to observations of "secondary causes" operating in nature, and
insisting that no conclusions be drawn beyond the evidence provided by
such observations, speculations of a cosmological sort were beyond its
scope. This conception of science therefore dictated a separation of
scientific conclusions from metaphysical speculation. Such a science
could come in handy to those, like Dawson, who owed much to the "twin
theologies" tradition. In violating the canons of Baconian method by
engaging in unwarranted speculation, Darwin (Dawson and Wilson alike
believed) had not only transcended the bounds of proper science, but
had obliterated the distinction between science and cosmology. As will
be shown, however, Wilson himself (and, for that matter, Dawson) violated the requirements of that Baconian tradition in order to engage in
his own cosmological speculations.

Like Dawson, Daniel Wilson was clearly disturbed by the events
that were then shaking modern science. The problem with Darwin's challenge to the supremacy of Baconian induction by his use of an hypothet-
ico-deductive method (which has since served as the methodological basis of modern evolutionary biology) was that it threatened the very

sense of mystery that provided the science of Wilson with its religious
and aesthetic dynamics. Not content to describe nature as it appeared
after the uncovering of each successive veil, Darwin had dared to assess the natural world as if no veils existed. His discovery of natural selection, an astute observer of the Darwinian method has noted,
"was, above all else, a triumph of reason. If banishing intuition from
our conception of the process of discovery deprives us of a sense of
mystery, it nonetheless permits us to analyze that process in a far
more satisfying manner than did the mythological accounts."[23] Such a
method may have been satisfying for the Darwinian, or may still be for
the modern biologist, but not for Dawson or Wilson. The Darwinian hypothesis, said Wilson, appears to have had as its immediate result
"the removal of many old land-marks of scientific faith, whereby we
witness some of the conditions of ruin, which mark all transitional
eras, - whether of thought or action. The old has been shaken, or
thrown down, the new is still to build."[24]

"In this light," Wilson's review of Darwin's *Origin* began, "...we
must look upon that comprehensive question which now challenges revision...*What is Species*?" Comprehensive and important because it "forces us back to first principles," and has equal effect upon paleontology, zoology, and the relations of science and theology, the species
question, as resolved by Darwin, stood as a challenge above all because
of its bearings upon Wilson's own major interest, ethnology. It involved not only the question "In what forms has creative power been
manifested in the succession of organic life?" but also another, more
central to Wilson's particular scientific concern: "Under what conditions has man been introduced into the most diverse and widely separated provinces of the animal world?" It is clear that to Wilson, the
ethnologist, the former question was in fact subsumed within the latter:

> It is to the comprehensive bearings of the latter in-
> deed, that the former owes its origin; for what is the
> use of entertaining the question, prematurely forced
> upon us: Are all men of one and the same species?
> While authorities in science are still so much at
> variance as to what species really is; and writers
> who turn with incredulous contempt from the idea that
> all men are descended from Adam, can nevertheless
> look with complacency on their probable descent from
> apes![25]

Much of Wilson's review of Darwin's *Origin* dwelt upon what most
observers agree to have been the fundamental challenge posed by Darwin
and he was able to summarize the essence of that challenge with admir-
able brevity:

> He has arrived at the conclusion that there is in
> reality no essential distinction between individual
> differences, varieties, and species. The well-marked
> variety is an incipient species; and by the operation
> of various simple physical causes and comparatively
> slight organic changes, producing a tendency towards
> increase in one direction of variation, and arrest-
> ment, and ultimate extinction in another, the law of
> *natural selection*, as Darwin terms it, results, which
> leads to his "preservation of favoured races in the
> struggle for life."[26]

When he rejected the "transmutation" theory of evolution through
"the tendency of species to an infinite multiplication of intermediate
links," Wilson also rejected the notion of natural selection. He noted
that the evidence of paleontology afforded no positive evidence to affirm the idea of transmutation. "We look in vain," he said, "among organic fossils for any such gradations of form as even to suggest a process of transmutation. Above all, in relation to man, no fossil form
adds a single link to fill up the wide interval between him and the
most anthropoid of inferior animals." This, he contended, is true even
if one limited oneself to the purely physical characteristics by which
the paleontologist is bound. But once one considered mental characteristics, as the ethnologist must, the challenge to Darwin's theory was
even stronger, for

> if the difference between man and the inferior ani-
> mals, not only in mere physical organization, but
> still more in all the higher attributes of animal
> life, be not relative but absolute, then no multi-
> plication of intermediate links can lessen the ob-
> stacles to transmutation. One true antidote there-
> fore to such a doctrine, and to the consequent de-
> nial of primary distinctions of species, seems to
> offer itself in such broad and unmistakable lines
> of demarkation [sic] as Professor Owen indicates,
> between the cerebral structure of man and that of
> the most highly developed of anthropoid or other
> animals.[27]

Here was the genesis of Wilson's argument a dozen or so years later
that Shakespeare's Caliban could be viewed as the "missing link." But
unlike the "missing links" so long sought by later anthropologists,
Wilson's did not serve the purpose of establishing once and for all
time the strength of the chain of evolution. It was, instead, a means
of establishing a kind of buffer between animal and man in creation.

Although Wilson rejected Darwin's theory, he treated the views of
the British naturalist with the respect due to any who seriously proposed a significant advance in scientific truth. "His 'Origin of
Species,' is no product of a rash theorist, but the result of the

patient observation and laborious experiments of a highly gifted naturalist." There could be little doubt that Darwin's work would create as
many problems as it solved. It would, Wilson noted, "tend to...give
courage to other assailants of those views of the permanency of species,
which have seemed so indispensable alike to all our preconceived ideas
in natural science, and to our interpretations of revealed cosmology."
In the end, Wilson said, no serious harm could come of Darwin's work:
if it was incorrect, scientists would prove it so; if correct, new
areas of knowledge would be opened. In either case the cause of Truth,
through the practice of science, would be served: "our attitude ought
clearly to be that of candid and impartial jurors. We must examine
for ourselves, not reject, the evidence thus honestly given....All
truth is of God, alike in relation to the natural and the moral laws,
and of the former, as truly as of the latter may we say: 'if this
counsel or this work be of men, it will come to nought; but if it be of
God, ye cannot overthrow it; lest haply ye be found even to fight a-
gainst God.'"[28]

Daniel Wilson's science was not, therefore, tied to a dogmatic
commitment to the Mosaic cosmogony; yet in its own way it was still
tied to the notion of a providential design and harmony in nature.
This resulted, at times, in an apparent abandonment of the essence of
scientific enquiry. On the one hand, Wilson applauded the work of the
comparative anatomist Richard Owen (who embraced evolution but rejected
natural selection) for his "grand generalizations, based not [as with
Darwin] on theory, but on laborious and exhaustive induction, [thereby
revealing] to us the plan of the Creator."[29] Yet on the other, he admitted that science - even that which is based upon the solid Baconian
observation of laws operating in nature - becomes inadequate. "Science
has achieved wondrous triumphs," he observed in his 1861 presidential
address to the Canadian Institute, "but life is a thing it can neither
create nor account for, by mere physics. Nor can we assume even that
the whole law of life can be embraced within the process of induction,
as carried out by an observer so limited as man is....Darwin, indeed,
builds largely upon hypotheses constructed to supply the gaps in the
geological records; but whilst welcoming every new truth which enlarges
our conception of the cosmic unity, all nature still says as plainly
to us...: 'Canst thou by searching find out the Almighty to perfec-
tion?'"[30] The reasons for Wilson's fascination with the natural world,
protected by the veils of Isis, ultimately lay in the satisfaction he

derived from the certainty that, behind each raised veil, was fortunately another which yet preserved the mysteries of life.

III

The appearance of Darwin's long-awaited sequel to *The Origin of Species* inaugurated a new wave of anti-Darwinian sentiment in Canada. By the end of the 1860s, little effective accommodation had been reached in Canada between the defenders of Christianity and the scientific findings of the decade. Articles recounting leading scientific infidelities continued to find a welcome place in the Methodist newspaper, the *Christian Guardian*,[31] as they had throughout the decade of the 1860s. Few Canadians writing for the *Guardian* were willing to accept Darwin's theory of evolution, and virtually none gave countenance to the idea of natural selection.

But with the publication of *The Descent of Man*, Canadian critics of modern science were drawn into a renewed struggle with the latest phase of the Darwinian heresy. Just after the publication of Darwin's new book, but before its contents were known in Canada, a long article entitled "Darwinism and Christianity" appeared in the *Christian Guardian*. It was well reasoned and moderate in tone, stating all the major objections of the decade to the theory of evolution and concluding that "all observation and research extending over the period of human and animal life testify to the permanence of species."[32] It was the last of such articles to speak with moderation of Darwin, for, within months, copies of *The Descent of Man*, stowed in the holds of steam packets plying the North Atlantic, reached Canadian shores. Not a few Canadians may later have wished that the ships carrying the offending cargo had met with a watery fate on the shoals of the St. Lawrence.

It was a book that focused the full force of evolutionary theory upon the question of the origins of man, a subject not mentioned by Darwin in *The Origin of Species*. Human beings, Darwin now argued, probably descended from some form of primate. He further enraged the Victorian sensibility by stressing the significance of "sexual selection," the choice of a reproductive partner, in the process of evolution. Finally, he argued - and here he challenged conventional mental and moral philosophy - that the intellectual and social faculties of man were adaptive and that his survival or extinction rested upon such changes. Darwin's new book involved not only the study of nature but

of human nature. *The Descent of Man* raised questions of fundamental
importance in the areas of psychology and philosophy, as well as geolo-
gy and anthropology. *The Origin of Species* had led men into the new
world of geological time; *The Descent of Man* took them on a journey into the interior of the mind.

In the dozen years between 1859 and 1871, Daniel Wilson, in Toronto, had followed the debate over Darwinism and had continued his ethno-
logical researches and writing. As professor of English literature as
well as history, however, he had also taught the works of Shakespeare.
Caliban: The Missing Link [1873], Wilson's response to *The Descent of
Man*, was thus a mixture of ethnology and literary criticism. It also
contained not a little sheer fancy. In 1887 G. Mercer Adam wrote an
article for *The Week* on Wilson. In it he described Wilson's book perfectly. *"Caliban,"* he said, "is an interesting Shakespearean study,
combining great imaginative power with a strong critical faculty, and
giving the reader much curious information, with not a little fanciful
disquisition, on the Evolution theory."[33]

It was, however else one might judge it, an extraordinary book.
All the more remarkable was the fact that although the book was based
upon Wilson's experience teaching Shakespeare's play *The Tempest* to
students of University College, his students doubtless came away from
their classes knowing at least as much about Darwin, evolution, and the
descent of man as they did about Prospero, drama, and the stuff of
dreams. "The leading purpose of the following pages," Wilson wrote in
his preface, was "...to shew that [Shakespeare's] genius had already
created for us the ideal of that imaginary intermediate being, between
the true brute and man, which, if the new theory of descent from crud-
est animal organisms be true, was our predecessor and precursor in the
inheritance of the world of humanity....A comparison between this Caliban of Shakespeare's creation, and the so-called 'brute-progenitor of
man' of our latest school of science, has proved replete with interest
and instruction to the writer's own mind; and the results are embodied
in the following pages."[34]

By the 1870s, Daniel Wilson - aided by his poetic imagination -
had left the more certain domain of his scientific expertise to consider the philosophical and psychological implications made necessary by
The Descent of Man. If one assumed that matter existed eternally, then
it was difficult to accept the coming into being (and, hence, existence)
of a Divine Creator. Was it not "more scientific," Wilson suggested,

"to start with the preoccupation of the mighty void with the Eternal
Mind"? But if (the Divine) mind existed before matter, how could this
be reconciled with the findings of evolutionary science, which implied
that mind was created through "the development of the intellectual,
moral, and spiritual elements of man, through the same natural selection by which his physical evolution is traced, step by step, from the
very lowest organic forms"?[35]

The scientific imagination had signally failed to resolve this dilemma; nor had scientists found any fossil evidence that could help.
Yet what science had failed to do, "the creative fancy of the true poet,
working within its own legitimate sphere, has accomplished to better
purpose." The "seductive hypotheses" and the "severer inductions" of
science, Wilson believed, inhibited a solution to the problem of the
descent of man. Fortunately, where Baconian inductions had proved inadequate, the poet had succeeded. Shakespeare himself had, in *The
Tempest*, "presented...the vivid conception of 'that amphibious piece
between corporeal and spiritual essence' [the quotation is from Sir
Thomas Browne's *Religio Medici*], by which, according to modern hypothesis, the human mind is conjoined in nature and origin with the very
lowest forms of vital organism. The greatest of poets...has thus left
for us materials not without their value in discussing...the imaginary
perfectibility of the irrational brute; the imaginable degradation of
rational man."[36]

Those last dozen words contain the key to Wilson's interpretation
of Caliban as the "missing link" between animal and man, as well as the
reason why he felt obliged to rely upon the poetic imagination in order
to bolster his own arguments against evolutionary science. It was the
mind of both animal and man which was under consideration, and it was
poetic insight, rather than scientific thought, that alone could provide the light necessary for the journey into the interior. "History
tells of the acts, literature tells of the mind," read the literature
notes of one of Wilson's students in the class of 1874.[37] Furthermore,
lacking what Wilson had earlier called the "well-regulated fancy" of
the poet, the scientist's vision was limited: he could conceive as
scientifically "imaginable" the degradation of the rational man, but
could only see as "imaginary" the possibility that the "irrational
brute" might develop - that is, might evolve within certain specific
limits - to the height of his potential and yet not become man.

For Wilson, Shakespeare's "hag-born whelp" was just such a creature: "the highest development of 'the beast that wants recourse of
reason.' He has attained to all the maturity his nature admits of, and
so is perfect as the study of a living creature distinct from, yet next
in order below the level of humanity."[38] Wilson's book sought in a
variety of ways to make the case for the inviolability and sanctity of
the gap between Caliban, the poet's imaginative conception of the "natural brute mind" developed to its fullest capacity, and even the most
degraded of rational human beings.

It was necessary first to dispose of the view that since scientific evidence indisputably showed that man was structurally far more similar to the higher apes than apes were to the lesser quadrumana, he was
evolved from the ape. The parallel, Wilson argued, was a false one.
Man and the apes are both animals; hence, they are physically similar.
Yet it does not necessarily follow that the former evolved from the
latter. Darwin had made this mistake in logic, and the result was of
immense import for Wilson's thought. It made the journey into the interior both possible and necessary. "To all appearance," he wrote of
The Descent of Man, "the further process in the assumed descent...of
man from the purely animal to the rational and intellectual stage, is
but a question of brain development; and this cerebral growth is the
assigned source of the manward progress: not the result of any functional harmonising of mind and brain....It is difficult to dissociate
from such an idea the further conclusion, that reason and mind are no
more than the action of the enlarged mind; yet this is not necessarily
implied....The brain is certainly the organ of reasoning, the vital instrument through which the mind acts; but it need not therefore be assumed that brain and mind are one." Caliban, for Wilson, marked the
symbolic midpoint between the brain of animals and the mind of man, the
narrow gulf between irrational ape and rational man: "No being of all
...Shakespeare[an] drama more thoroughly suggests the idea of a pure
creation of the poetic fancy than Caliban. He has a nature of his own
essentially distinct from the human beings with whom he is brought into
contact. He seems indeed the half-human link between the brute and
man; and realises, as no degraded Bushman or Australian savage can do,
a conceivable intermediate stage of the anthropomorphous existence, as
far above the most highly organized ape as it falls short of rational
humanity."[39]

Wilson was not content to rest after writing about "The Monster Caliban." His next chapter was entitled "Caliban, the Metaphysician." Here he attempted to rebut the evolutionist's claim that all the attributes of humanity - man's "intellect, his conscience, and his religious beliefs" - were simply part of the evolutionary process. As with Dawson, this was Wilson's central source of concern over the question of man's descent. "The growing difficulty, indeed," he wrote when considering the possibility of the evolution of these attributes, "is not so much to find man's place in nature, as to find any place left for mind." Nor was this the only problem. "If conscience, religion, the apprehension of truth, the belief in God and immortality, are all no more than developed or transformed animal sensations; and intellect is only the latest elaboration of the perceptions: it need not surprise us that inquiry has already been extended in search of relations between the inorganic and the organic. On this new hypothesis of evolution 'what a piece of work is man!' and as for God, it is hard to see what is left for Him to do in the universe."[40]

It is not surprising, therefore, that Wilson rejected the interpretation of Caliban rendered by Robert Browning in his poem "Caliban upon Setebos." The thoughts of Browning's Caliban suggested to Wilson that poor Caliban, after all "only a poor half-witted brute, - [had got] terribly out of his depth." Shakespeare's Caliban, and Wilson's, had been a poetic rendering of the "intermediate, half-brute, missing link." Browning's was that of the Darwinians: "the human savage, grovelling before the Manitou of his own conception."[41]

The essential points have now been made. Wilson responded strenuously and at some length to the challenges posed by evolutionary science to his scientific and religious beliefs. He replied to Darwin from the perspective of his own scientific expertise, but in the course of the dozen years between the publication of *The Origin of Species* and *The Descent of Man* two things happened. First, he abandoned the premises of science in important ways in order to defend the religious heritage he saw under attack, and he did so by returning to the poetic fancy with which he had always been enamoured. Many Canadians, whether university students, clergymen, or Victorian families, perplexed by what they read in the *Christian Guardian* or other newspapers on the subject of evolution, may well have found the writings of Wilson to be the proper response for Christians to make to such heresies; but it is questionable whether to the more reflective among them these

writings assuaged the nagging uncertainties that were in the air by the
decade of the seventies.

The second observation to be made is that whenever Wilson considered the place of man within the framework of the evolutionary hypothesis,
he was forced to deal with questions of immense philosophical import.
Darwin's new book called fundamentally into question the various dual-
isms - spiritual and material, revelation and nature, mind and brain -
which were basic to the science and the religion of Wilson and others.
It marked the bankruptcy of the Baconian scientific method, which actively avoided hypothetical assertions about the existence of natural
laws. Furthermore, it called into question the mechanistic faculty psychology on which their conceptions of the human constitution in its mental aspects were based. "The nineteenth century...has...failed as yet
to arrive at a satisfactory conclusion as to man's place in nature,"
wrote another Canadian scientist in 1872. "We think," he later stated,

> ...that any naturalist is justified, as a scientific
> man, in maintaining that all classifications of man
> by his anatomical characters alone, are *artificial*,
> and as such are indefensible. Such classifications
> do not embrace the totality of man's organization,
> and can not, therefore, be natural....man's zoolo-
> gical definition must be made to include something
> more than his mere physical and anatomical struc-
> ture. *That* something is man's mental and moral
> constitution; and we repeat our belief that any
> naturalist is justified, without disparagement to
> either his knowledge or his ability, in maintain-
> ing that man's psychical peculiarities are as much
> an integral factor of his zoological definition as
> his physical structure, or perhaps more so.[42]

Such concerns for the legitimacy of the "mental and moral constitution" and for man's "psychical peculiarities" may well have been within the scientist's area of inquiry, but they were also laden with philosophical implications which transcended the limits of natural science.
The high-water mark of the public influence of the earth sciences had
been reached by the 1870s. Thereafter, in an ever-increasing degree,
the educated public would look to social philosophers attuned to the
new physiological sciences for solutions to the problems posed by modern thought. And by the 1870s, discussions of the nature of the human
condition could also take place in forums other than the cloistered
confines of Canadian educational institutions and the pages of religious newspapers. The ranks of orthodoxy had been breached, not only
by new ideas but also by the fact that members of the lay public could

now put forward their considered views in new magazines of informed opinion given their start by strange winds that gusted in the new Canadian nation.

In the end, as we reflect upon the place of Daniel Wilson in the Darwinian debate, we are reminded of one of the better-known aphorisms of a nineteenth century social philosopher whose stock was never to be very high in Canada. "The ultimate result of shielding men from the effects of folly," Herbert Spencer once wrote, "is to fill the world with fools."[43]

FOOTNOTES

1   George Basalla, William Coleman, and Robert H. Kargon, eds., *Victorian Science: A Self-Portrait from the Presidential Addresses to the British Association for the Advancement of Science* (New York, 1970), 48.

2   "British Association for the Advancement of Science," *Canadian Journal* [*CJ*], ser. 2, V (Mar. 1860), 66-7.

3   *Ibid.*, 67-8.

4   Basalla et al., *Victorian Science*, 19-20, 132, 400, 412-13, 415-16; George H. Daniels, *Science in American Society* (New York, 1971), 45-6; George H. Daniels, *American Science in the Age of Jackson* (New York, 1968).

5   See Daniels, *American Science*, 66, for a short but apt description of the "Baconian method".

6   Francis Bacon, *The Works of Francis Bacon*, ed. James Spedding et al. (London, 1883), IV, 32-3. I am indebted to Dr. Graham Reynolds for this reference.

7   Daniel Wilson, "The President's Address" (Read before the Canadian Institute, 7 Jan. 1860), [*CJ*], ser. 2, V (Mar. 1860), 127.

8   He had also written pieces for *Tait's Magazine*, *Chambers' Information for the People*, and *The Scotsman*, among others. See Langton, *Wilson*, 36-9. Further information on Wilson may be obtained from these sources: H. H. Langton, *Sir Daniel Wilson: A Memoir* (Toronto, 1929); Jessie Aitken Wilson, *Memoir of George Wilson* (Edinburgh, 1860). (George Wilson was Daniel Wilson's young brother, who was Regius Professor of Technology at Edinburgh University and director of the Industrial Museum of Scotland until his death in 1859); G. M. Adam, "Daniel Wilson," *The Week* IV (6 Oct. 1887), 726-7; T. F. McIlwraith, "Sir Daniel Wilson: A Canadian Anthropologist of One Hundred Years Ago," *Transactions of the Royal Society of Canada*, sec. II, ser. 4 (1964), 129-36. Bruce G. Trigger, "Sir Daniel Wilson: Canada's First Anthropologist," *Anthropologica*, n. s. VIII (1966), 3-28.

9   Daniel Wilson, *Prehistoric Annals of Scotland*, 2nd ed. (London 1863), I, xvii.

10 *Ibid.*, 119.

11 Daniel Wilson, "Sonnet--from 'Spring Wild Flowers,'" *Canadian Monthly and National Review* VIII (July 1875), 8.

12 Quoted in a review (anonymous) of *Prehistoric Man: Researches into the Origin of Civilization in the Old and New Worlds*, in *British American Magazine* I (1863), 92. The fact that the reviewer chose this passage from the many other quotable ones in Wilson's book

illustrates the hold of the metaphor of the veil upon at least one person other than Wilson and myself.

13   James Bovell, *Outlines of Natural Theology, for the Use of the Canadian Student* (Toronto, 1859), iii.

14   Basalla et al., *Victorian Science*, 48, 53.

15   Editorial, *Dalhousie Gazette* IV (27 Apr. 1872), 10-11.

16   John C. Greene, *Darwin and the Modern World View* (Baton Rouge, 1973), 4 *passim*. The historiography of nineteenth-century science is both brilliant and fascinating. See, for example, Loren Eiseley, *Darwin's Century* (New York, 1961); John C. Greene, *The Death of Adam* (Ames, Iowa, 1959); Cecil J. Schneer, *Mind and Matter* (New York, 1970).

17   G. Mercer Adam, "Prominent Canadians. II. Daniel Wilson," *The Week* IV (6 Oct. 1887), 727.

18   William Nesbitt Ponton, "Lecture Notes--Professor Wilson's Lectures--1873--1st year," notebook, pp. 13, 15; "Rhetoric," 63-4, University of Toronto Archives, University of Toronto. Emphasis in original.

19   Wilson, "President's Address" (1860), 115-16.

20   See Michael T. Ghiselin, *The Triumph of the Darwinian Method* (Berkeley, 1972). Ghiselin provides a remarkably brief but accurate description of the theory of natural selection as put forward by Darwin: "Organisms differ from one another. They produce more young than the available resources can sustain. Those best suited to survive pass on the expedient properties to their offspring, while inferior forms are eliminated. Subsequent generations therefore are more like the better adapted ancestors, and the result is a gradual modification, or evolution. Thus the cause of evolutionary adaptation is differential reproductive success" (46).

21   The very characteristics rejected by Wilson constituted the essence of the Darwinian "revolution" in scientific method. See *ibid.*, 63. It must be stressed that these reactions by Dawson and Wilson were by no means unique or uncommon. Similar objections were voiced by the most prominent British and American scientists of the day. See Sir William Armstrong's presidential address to the British Association for the Advancement of Science: "But when natural selection is adduced as a cause adequate to explain the production of a new organ not provided for in original creation, the hypothesis must appear to common apprehensions, to be pushed beyond the limits of reasonable conjecture" (reproduced in *CJ*, ser. 2, IX [1864], 37). See also the presidential address several years later to the American Association for the Advancement of Science by J. Lawrence Smith. Darwin, said Smith, "is to be regarded more as a metaphysician with a highly wrought imagination than as a scientist....He is not satisfied to leave the laws of life where he finds them, or to pursue their study by logical and inductive reasoning. His method of reasoning will not allow him to remain at rest; he must be moving onward in his unification of the universe" (reproduced in the *Canadian Naturalist and Quarterly Journal of Science*, n. s. VII [1875], 148-55.

22 J. W. Dawson, "Review of 'Darwin on the Origin of Species By Natural Selection,'" *Canadian Naturalist and Geologist* V (1860), 100.

23 Ghiselin, *Darwinian Method*, 63, 76-7.

24 Wilson, "President's Address" (1860), 116.

25 *Ibid.*, 116, 117. Emphasis in original.

26 *Ibid.*, 118. Emphasis in original.

27 *Ibid.*, 120, 121.

28 *Ibid.*, 119.

29 Wilson, "President's Address" (1860), 122.

30 Wilson, "President's Address" (1861), 114.

31 See for example W. H. W., "Darwinism and Christianity," *Christian Guardian* XLI (28 Dec. 1870), 204.

32 "Darwinism and Christianity."

33 G. Mercer Adam, "Daniel Wilson," *The Week* IV (6 Oct. 1887), 726.

34 Daniel Wilson, *Caliban: The Missing Link* (London, 1873), xi-xii.

35 *Ibid.*, 2-4.

36 *Ibid.*, 8, 11.

37 Ponton, "Lecture Notes," p. 26: "Literature," University of Toronto Archives.

38 Wilson, *Caliban*, 78.

39 *Ibid.*, 20-1, 90-1.

40 *Ibid.*, 93.

41 *Ibid.*, 101, 113.

42 H. Alleyne Nicholson, "Man's Place in Nature," *Canadian Monthly and National Review* I (Jan. 1872), 35-7.

43 Quoted in Tibor R. Machan, *The Pseudo-Science of B. F. Skinner* (New Rochelle, N. Y., 1974). A more thorough discussion of the Darwinian "debate" in Canada may be found in A. B. McKillop, *A Disciplined Intelligence: Critical Inquiry and Canadian Thought in the Victorian Era* (Montreal: McGill-Queen's University Press, 1979).

III. SOCIAL DIMENSIONS OF SCIENCE AND PSEUDO-SCIENCE

The impact which science has on society has been of interest and
concern to perceptive individuals from the advent of heliocentrism and
natural selection through nuclear fission and recombinant DNA. The
writings of Robert Merton and Thomas Kuhn have led to a still burgeon-
ing interest in the reciprocal influences which society has on science.
Given the difficulties inherent in providing science/pseudo-science
demarcation criteria, it would appear obvious that there should be
similar interest in the interrelationships between pseudo-science and
society. One's first impulse is to think of these influences as being
negative. Since pseudo-science is in some sense "false" science, its
role in society must be a misleading one. Since society's interests
lie in producing valid sciences as sources of knowledge and control,
its role in pseudo-science must be a repressive one based on fairly
objective criteria of truth. But is this really the case? What role
do social, political and economic forces external to science proper
play in attempting to establish the science/pseudo-science demarcation
and in the support or repression of the resulting substantive areas of
knowledge or pseudo-science? What reciprocal role do the sciences and
the pseudo-sciences play in establishing social arguments?

James Jacob has taken a new look at an old theme, the relationship
between religion and science in seventeenth-century England. His focus
is the attack by Henry Stubbe on the writings of Sprat and Glanville in
particular and on the Royal Society in general. Stubbe's defense of
Aristotelianism as a bulwark against popery and his castigation of the
Royal Society as an institution inimical to Anglicanism and the monarchy is read by Jacob as philosophical and religious radicalism, not
as the conservatism it has traditionally been argued to be. He sees
Stubbe's reaction against the Royal Society and the new philosophy for
which it stood as having been based not on Stubbe's fear that it would
undermine traditional Christianity but, rather, on the belief that it
supported a debased and superstitious Christianity. Or, to put it in
Jacob's own words, that Stubbe "defended a backward science for the
sake of a radical theology." This thesis he sees as calling into question the usefulness of the distinction between Restoration science and
pseudo-science. A paper by Margaret and James Jacob "The Anglican

Origins of Modern Science: The Metaphysical Foundation of the Whig
Constitution" provided another "social" view of the development of
science in seventeenth-century England, this time emphasizing the reciprocal influences of the new philosophy and Whig constitutionalism.
In this paper, which was read at the conference but is being published
elsewhere, they discussed attempts to use science as a tool in lessen-
ing the radicalism of the English Revolution and argued that important
ideological support was provided to the Protestant monarchy through
scientific arguments. Among their conclusions was the interesting,
and somewhat unconventional observation that "the triumph of Newtonian
science represents another victory for the Whig constitution."

Roger Cooter moves away from the seventeenth century and, using
nineteenth-century phrenology as his focus, looks at the social and
political function of the term 'pseudo-science.' He concludes that
"the label 'pseudoscience' has played an ideologically conservative
and morally prescriptive social role," and that the function of the
label has been that of "a process of conserving social interests."
Beginning with the argument that 'pseudo-science' plays a conservative
role even when used in support of liberal arguments, Cooter then turns
to his analysis of phrenology and its social/political ideology. He
strives to clarify not only the role of those ideational systems termed pseudo-sciences, but also the social systems which apply the label.

Margaret Osler focuses her attention on the attempts of social
and political apologists to use the findings of science to support non-
scientific social arguments. Beginning with social Darwinism and
pseudo-physical arguments against the education of women in the nineteenth century, she presents the concept of the naturalistic fallacy
as one basis for her discussion. Denying that science is value-free,
Osler comments on the influence of the prevailing society's values on
science and suggests that an objectivity of sorts may be assumed if
the values contained within the science are independent of the social
thesis to be supported. Returning to an analysis of the naturalistic
fallacy, she concludes that, "in any attempt to derive an 'ought' from
an 'is', either the attempt fails outright, or the 'is'-statement already contains a hidden 'ought'-statement," and that since several
incompatible conclusions may follow from the same set of premises,
naturalistic systems of ethics "utterly fail to provide a scientific
basis for ethics."

ARISTOTLE AND THE NEW PHILOSOPHY:
STUBBE VERSUS THE ROYAL SOCIETY

This is an essay in historical demolition and reconstruction.

The Royal Society, founded in 1660 and chartered two years later, was dedicated to promoting a certain kind of science, the experimental or corpuscular philosophy. In the course of the 1660s various apologies for the Society and its Baconian program appeared - among the most important of them Thomas Sprat's *History of the Royal Society* (1667) and Joseph Glanvill's *Plus ultra: or, the progress and advancement of knowledge since the days of Aristotle* (1668). As the latter title indicates, these works by Sprat and Glanvill explored a common theme, the progress of knowledge and particularly the superiority of the new experimental philosophy of the Royal Society to the Aristotelian philosophy of the past. Glanvill and Sprat agreed that the experimental program of the Society represented a surer way to the discovery of useful and practical knowledge than the traditional method based upon the philosophy of Aristotle, adopted in the Middle Ages and only recently displaced by Bacon, Boyle, and the experimentalists of the Royal Society. Sprat went even further and argued that the experimental philosophy was not only basic to the pursuit of science but also provided the foundations for true religion, domestic peace and national wealth and power. It was a formula at once applicable to the reform of science and society.

The patient, cautious, industrious inquiry required in the conduct of experimental science would have a ripple effect in society at large. Most of all the pursuit of science would enhance national wealth and power. It would also create new industries and hence increase employment, thus reducing the number of the idle poor. To the extent that men were set to work, the threat of subversion from below would be overcome, and men would bury their differences in favor of the economic opportunities that science creates. Finally, the scientific temperament, modest, humble and skeptical, would produce good manners and foster true Protestant Christianity, to Sprat's mind identical with Anglicanism, because modesty and humility would cure sectarian enthusiasm and cautious doubt would prevent men from giving wholesale submission to papal authority. Here was Sprat's version, masterminded by John

Wilkins, of the natural religion of the Royal Society - liberal, tolerant, moderate, and visionary, millenarian even, in the sense that experimental science was seen as offering a path to Jerusalem.[1]

Regardless of its obvious and carefully contrived appeal, this
vision did not go unchallenged. As soon as Sprat and Glanvill had published their apologies the reaction set in, and from one quarter in
particular, the pen of Dr. Henry Stubbe, a West Country physician. It
is not clear, and may never be, what provoked Stubbe to enter the fray
and deliver the most violent and publicized attack on the Royal Society
that it has ever incurred. Stubbe himself claimed that he was stirred
to write his attacks because Glanvill and Sprat had impugned "the Aristotelian and Galenical way" of practicing medicine on which his own
considerable reputation as a physician rested.[2] But he was accused by
his contemporaries of being paid to write on behalf of others who could
thereby safely conceal their identities while fomenting opposition to
the Society. Among those accused of hiring Stubbe were the Royal College of Physicians, the Apothecaries, the clergy in the universities
and the aged Aristotelian Robert Crosse, who was already engaged in a
polemic against Glanvill in Somerset where Glanvill was Rector of Bath
and Crosse Rector of Chew Magna in the same neighborhood.[3] Certainly
Stubbe initially took up the Aristotelian cause by coming to Crosse's
defense against Glanvill in 1668, the year before Stubbe chose to put
anything in print. He practiced medicine during the summer season in
Bath and Bristol and caused a considerable stir throughout rural Somerset by defending Crosse and Aristotle against Glanvill and the Royal
Society.[4]

Whether we shall ever know why precisely Stubbe initiated his attacks, he soon found his stride, developed a standard body of arguments
and wrote and published, with the help of London supporters, a half
dozen or so tracts expounding his position.[5]

Stubbe's argument was pitched on two grounds. First, contrary to
Glanvill, he claimed that Aristotelian natural philosophy, particularly
when applied to medicine, far from being bankrupt, was superior to the
so-called experimental philosophy of the Royal Society. Glanvill's
deprecation of Aristotle, Stubbe claimed, was due to his ignorance of
ancient philosophy rather than to the superiority of the moderns.[6] But
he maintained that there was a more serious case to be made against
Glanvill and Sprat, and this led him to his second argument.

Not only was the comparison between the ancients and the moderns invidious and hence harmful to medical science. To the extent that such a view took hold and displaced the traditional scholastic curriculum in the schools and universities, it would prove the undoing of church and monarchy. Specifically, England would fall prey to the heretics, especially papists, Socinians and atheists, and Stubbe's worst and principal fear was that the papists would win out and popery, the embodiment of Antichrist, prevail.[7] Stubbe derived his argument for the potential triumph of popery from the strategy of the early seventeenth century magician Tommaso Campanella, for the re-Catholicization of Protestant Europe, especially England and Holland. Campanella claimed that Catholicism might be reintroduced via natural philosophy which would divert men's attention from religion and allow the papists to intrude their faith among men thus preoccupied with science.[8]

Stubbe invoked Campanella's strategy because he claimed that Sprat's *History* would have the same effect, though in print at least he left open the question of whether or not this was Sprat's intention. Sprat argued in *The History* that the Royal Society should accept to membership men of all religious communions, including Catholics, because it was not interested in what divided men but in scientific collaboration which in the long run in any case would lead them to bury their differences and live together in peace and prosperity.[9] In reply to Sprat's argument Stubbe writes:

> ...what benefit and advantage popery may derive from this, that our nobility and gentry, our divines and laity, laying aside all memory of the French and Irish massacre, and Marian persecutions, the gunpowder treason, the firing of London, and forgetting all animosities and apprehensions of future dangers, converse freely with and write obligingly to them [i.e., Catholics], testify a great esteem of them, and from the disuse of all harsh but too true censures, come at length to lay aside all rancor...; I say, how great benefit popery may draw hence, I cannot well comprehend:...[10]

Stubbe was claiming that in effect if not intentionally Sprat's *History* represented Campanella's scheme revived: "...how would Campanella have clapped his hands for joy to see this happy establishment [the Royal Society] which he so long ago projected in order to the converting of England, Holland, and other heretical countries?" Stubbe asks rhetorically.[11]

The danger posed by the experimental philosophy of the Society was not limited to science and medicine. Knowledge and learning, Stubbe

and Sprat agreed, played a vital role in determining the character and
fate of religion and polity, and Sprat's formulas, according to Stubbe,
would undermine both in favor of popery. The natural religion of the
Royal Society must not be allowed to prevail. The traditional Aristotelian curriculum must be preserved in the face of it because, Stubbe
argued, Scholastic philosophy not only provided the best medical knowledge; it also protected church and state against their enemies, papists, atheists and Socinians. To quote Stubbe:

> ...the politics of Aristotle suit admirably with
> our monarchy....The ethics...[of Aristotle]
> are generous, and subservient to religion, and
> civil prudence, and all manner of virtue: the
> logic and metaphysics [of Aristotle] are...entwisted
> with the established religion, and...requisite to
> the support of it against Papists and Socinians
> ....[12]

Stubbe then presented himself as being opposed to Sprat and Glanvill's
reading of the mission of the Royal Society because they would subvert
church and monarchy; he also presented himself as being equally committed to the philosophy of Aristotle for the support it gave to medicine, established religion and government.

On this basis Stubbe has been uniformly interpreted by scholars as
a conservative defender of the status quo in science, learning and religion. Herschel Baker in an influential book published in 1947 and
R. F. Jones in his classic study, *Ancients and Moderns*, are chiefly responsible for inventing this interpretation. Baker describes Stubbe as
"that irascible and unlovely conservative...seeking to check the tide
of science and 'progress.'"[13] Jones for his part speaks of Stubbe's
"conservative spirit, with its genuine fear of change."[14]

I intend in the remainder of this paper to challenge this interpretation by offering a new reading of the evidence, a reading which, instead of making Stubbe a conservative, renders him into precisely the
opposite - a philosophical and religious radical - and sees a consistency between his career before 1660 and his polemical involvements after the Restoration.

Stubbe was a radical Independent and republican during the 1650s,
influenced by James Harrington, Sir Henry Vane and Thomas Hobbes. From
Harrington came his republicanism, from Vane his republican Independency.[15] From Hobbes came materialism. In fact Stubbe was so enthusiastic about Hobbes's ideas that he undertook, with the master's blessing,
to translate *Leviathan* into Latin (the project was never completed).[16]

The chief bond between Stubbe and Hobbes, as their surviving correspondence reveals, was their mutual opposition to the Presbyterians
and especially to the Presbyterian (and Roman Catholic) notion that
the clergy are ordained by God to serve as his representatives on earth
and thus possess spiritual authority which is independent of the authority of the civil sovereign. This view equally violated Hobbes's
notion of the exclusive authority, in church as well as state, of the
civil ruler and Stubbe's notion, deriving from his radical Independency, of every man's freedom to worship as he chose without clerical
intervention.[17] The Presbyterian conception of the role of the clergy
was also based upon the notion of a separate spiritual realm, the existence of which both Hobbes and Stubbe, under Hobbes's influence, rejected. For both of them the spiritual was rendered material.[18]
Stubbe then was a Hobbesist of sorts, though not completely because he
would have left everyone free to witness to the truth as the spirit
(interpreted in a material sense) moved him, whereas for Hobbes the
power to determine what was true belief and to enforce it belonged exclusively to the civil sovereign. Both were extreme Erastians, but
Stubbe was an Erastian libertarian.[19]

This was Stubbe in the late 1650s and he wrote tracts in which he
argued for his Erastian libertarian ideal as the basis for a religious
settlement. The bottom fell out in 1660 and Stubbe had to tread cau-
tiously in order to survive in a world where there was no room for re-
publicans and Independents. Stubbe conformed to the established church
in 1662, but he said he did so because it was "the most national"
church, which does not necessarily represent a return to orthodox religious views.[20] Indeed his Hobbesian or quasi-Hobbesian materialism or
naturalism comes to the surface again in 1666.

The particular issue at hand was the king's power to cure, known
as the royal touch. A gentleman named Valentine Greatrakes was going
about the countryside performing cures for scrofula, a disease known as
the king's evil becuase it was supposed to be especially susceptible of
treatment by the royal touch. And here was Greatrakes, a mere commoner
(though a gentleman), achieving the same results as the king by what
was called "stroking" - hence his reputation as "the Stroker" or "the
Irish Stroker" (he was an English landlord in Ireland).[21] The subversive implications of Greatrakes' success were lost on no one because
the king's healing powers were regularly invoked in the early

Restoration as signs that his authority came from God or was at least
thus attested to by God.[22]

In the midst of the furore Stubbe produced a tract, entitled *The
miraculous conformist*, defending the authenticity of Greatrakes' cures.
One detects the remnants of Stubbe's pre-Restoration radicalism and
republicanism in his argument in the tract that the power to heal is
inherent in nature, hence accessible to men like Greatrakes and not
limited to Christ, the Apostles or the king.[23] Stubbe was by implication taking a swipe at divine-right monarchy, which was accepted orthodoxy after 1660, by democratizing or at least gentrifying (Greatrakes
was a gentleman after all) the royal touch. He was also casting doubt
upon Christian revelation, confirmed by the apostolic power to perform
miracles, by naturalizing that power.

How then does one square Stubbe the heterodox critic of divine-
right monarchy in 1666 with Stubbe the defender of church and state
against the likes of Sprat and Glanvill three years later? There are
two possibilities. Either Stubbe changed his mind between 1666 and
1669 or his defense of church and state in 1669 is not what it appears
and has been interpreted to be. There is no evidence for the first
supposition - that he changed his mind. But there is considerable evidence for the second, and this entails reading his tracts against
Sprat and Glanvill more closely than before and then examining Glanvill's response to them.

When this closer reading is performed, a picture emerges that
throws considerable doubt on the standard interpretation of Stubbe's
attacks because it is revealed that, although he defends church and
monarchy, his view of both is so far from the accepted one that more
conventional apologists might well have wondered that with friends like
Stubbe, who needed enemies? Certainly if measured by the yardstick of
his own time, he is a long way from the conservatism that Baker and
Jones impute to him. Let us then take that yardstick and hold it up to
Stubbe's views, as these are revealed in the tracts on the Royal Society.

Writing in 1670 about why he undertook his attacks, he says:

> ...I had...powerful inducements...; and those are
> the exigencies of the English monarchy, whereunto...
> it is the prudence of every particular person to
> contribute all he can to the support of it, against
> all such intendments as may...introduce popery on
> the one side...and against all anarchical projects
> or democratical contrivances, whereof a debauched
> and ungenerous nation is not capable....[24]

Clearly he has discarded his republican commitments of the 1650s, and
equally clearly he sees popery to be a real threat, as many people did,
especially by 1670, when both the Queen and the Duke of York were known
to be devout Catholics. Stubbe is now committed to the restored Protestant monarchy as a defense against both popery and democracy. But
he pitches his argument on the secular grounds of prudence rather than
the religious grounds of duty and devotion to a king who rules by divine right. This latter was the official and conservative view at the
time, the view preached from Restoration pulpits not to be upset until
the events of 1688-89, whereas Stubbe's prudentialism would have sound-
ed suspiciously Hobbesian to a sensitive Restoration ear.[25] Hobbes's
works had put forward in bold form a *de facto* theory of sovereignty
which held that men subject themselves to their rulers not for religious reasons but out of prudent self-regard.[26] Fortunately, there is
contemporary confirmation of this Hobbesian reading of Stubbe's meaning. Glanvill detected Stubbe's prudentialism and attacked him for the
rejection implicit in it of the religious argument for obedience.[27]
Stubbe's prudentialism may also have been what Glanvill had in mind
when he associated Stubbe's views with Hobbes.[28]

Glanvill catches more than this in Stubbe's explanation of his
commitment to the restored monarchy. In the passage I have just quoted
Stubbe opposes "all anarchical projects or democratical contrivances,
whereof a debauched and ungenerous nation is not capable...." Glanvill
uses the passage as an example of Stubbe's duplicity. Stubbe condemns
the republican experiments of the 1650s of which he was so much a part,
but the fault lies with the people not with the republican ideal,
"whereof," as he says, "a debauched and ungenerous nation is not capable." As Glanvill points out, Stubbe here implies that a monarchy is
the government appropriate to "a debauched and ungenerous nation,"
which was certainly not the accepted view at the time.[29] It is rather
a view, like Stubbe's prudentialism, that sees politics as contingent
upon time and circumstance rather than as made in heaven, as relative,
not absolute, and this relativism ran in the face of religious conven-
tion; indeed it would have been regarded as subversive in the paranoid
atmosphere of the early Restoration.

Stubbe says other things about monarchy, which Glanvill does not
pick up, that would have given equal offense to official and pious
opinion. Against Sprat's views he rejects "paternal right" and "primo-
geniture" as grounds for "sovereignty," thus again raising the specter

of Hobbes.[30] After the publication of his principal attack on Sprat's
*History, A Censure upon Certain Passages contained in the History of
the Royal Society*, he is accused of having libelled the king.[31] What
was it in the *Censure* that provoked that response? One cannot be certain, but there is a section which must have raised more than a few
eyebrows. Stubbe accuses Sprat of justifying absolute monarchy or despotism[32] and goes on to state his own view of government, which gives
every subject the right to submit the commands of his earthly rulers to
the scrutiny of his conscience in order to determine whether or not
those commands violate the will of God.[33] What Stubbe in fact does is
to rehearse sixteenth-century resistance theory deriving from Calvin
and elaborated by his, chiefly Huguenot, interpreters. I know of no
other printed apology for this theory so early in the Restoration.[34] It
was not an apology calculated to please the authorities in 1671, and yet
it is slipped into a tract ostensibly written to defend the Restoration
settlement. Either Stubbe did not know what he was doing or he knew
all too well and was attempting to cover his tracks by saying one thing
and meaning another or by saying contradictory things at the same time
in such a way as to mask subversive meanings by professions of alle-
giance, in a word, by resorting to subterfuge. Whether Stubbe knew
what he was doing or not, his apology for resistance theory - and much
else - does not fit the standard account delivered by Baker and Jones.

        Stubbe's heterodoxy manifests itself in his religious views as
well as his political ones. True to say, he does argue that orthodox
Anglican doctrine is so bound up with Aristotelian logic and metaphysics that any departure from the latter in the direction of experimental
philosophy and what he calls "mechanical education" would jeopardize
the established church.[35] In particular such a departure would weaken
the church's ability to defend itself against attack from papists,
Socinians and atheists because the church would lack the clergy skilled
in the Scholastic mode of disputation needed to put up a good fight.[36]
But his argument is curiously left-handed. While claiming that religious education should be founded upon Aristotelian philosophy, he does
so because such "instruction would represent those ceremonies and habits of the church as decent, orderly and rational which would otherwise
seem uncouth and phantastical."[37] In other words, the bare bones of
established church doctrine might not be palatable without a mantle of
Aristotelian metaphysics thrown over them - not the message one would
expect to get from a defender of the orthodox faith. Nor is this all
the evidence that gives the lie to Stubbe's supposed orthodoxy.

In his *Censure* of Sprat's *History* Stubbe claimed that Sprat's view
of religion unwittingly supports the ancient Arian heresy which denies
Christ's divinity and hence the Trinity and which in its modern form
was known as Socinianism. Sprat had argued against the linkage between
Aristotelian philosophy and Christian theology by saying: "Religion
ought not to be the subject of disputations: It should not stand in
need of any devices of reason: ...nothing else is necessary but a bare
promulgation of Christian doctrine, a common apprehension, and sense
enough to understand the grammatical meaning of ordinary words."[38]
Stubbe seized upon this passage and maintained that the upshot of
Sprat's argument, were it to take effect, would be for Christianity to
become purely scriptural, which would be a victory for Socinianism, that
Sprat's *History*, in other words, by insisting that religion be severed
from metaphysics and allowed to rest on "the grammatical meaning of ordinary words," as found in the Bible, would give encouragement to the
Socinians.

Stubbe claimed to be defending the established church from Arians
and Socinians as well as papists by preserving Aristotle from the de-
predations of Sprat and Glanvill. But there is an implicit assumption
in his suggestion that Sprat's insistence upon a scriptural religion
based not upon Aristotelian philosophy but rather "the grammatical
meaning of ordinary words" in the Bible would hand religion over to the
Socinians. The assumption is that the Bible does not support the
weight of Aristotle imposed upon it, that on its own it is an Arian or
Socinian document, that in effect the doctrines of Christ's divinity
and the Trinity (not to say any others) are not biblical but Aristotelian, that orthodox religion is unscriptural and true (Arian) Christianity is heretical in orthodox terms. Once more Stubbe is pretending
to argue for orthodoxy but is in fact insinuating something - this time
the identification of scriptural religion with Socinianism - which
would leave orthodoxy aghast. Listen to the subterfuge in Stubbe's own
words:

> It is but too apparent that those in our days who
> join with the Arians in decrying new words and
> such as are not in Scripture, who think that
> Christianity ought not to be confined to any
> methodical creeds or articles...; 'tis manifest
> that they look with indifference on the things
> signified by those words and forms; 'tis mani-
> fest that they overthrow the constitutions of
> the Church of England, whose *Articles* make use

> of those significant terms, transmitted from the
> Fathers to our Schools; and subvert the basis of
> our religion, as it is represented in our laws.
> ...39

It is the repeated use of the phrase "'tis manifest" that gives Stubbe
away. Manifest to whom, the orthodox might well ask? Manifest to me,
Stubbe would reply, because he comes to the issue already convinced of
something no orthodox Christian would ever admit, namely, that the
"ordinary words" of the Bible do not support such doctrines as Christ's
divinity and the Trinity, that on its own terms, divested of the metaphysical superstructure raised upon it to sanction such doctrines, the
Bible is a Socinian document.

   Nor is this the only evidence indicating that Stubbe was himself
an Arian. In 1666, when he produced *The miraculous conformist*, arguing
against the king's monopoly of the so-called royal touch and against
the Christly and apostolic monopoly of power to perform biblical miracles, he had been accused of some form of Arianism. The accusation,
made in a letter from Daniel Coxe to Robert Boyle, is revealing. Coxe
had learned

> by them who have reason to pretend to understand
> Dr. Stubbe his designs...that Dr. S. intends to
> demonstrate from what Greaterex hath performed
> that the miracles of our blessed Savior were not
> derived from any extraordinary assistance of a
> divinity, much less from the union of the divine
> nature with humanity...

Coxe goes on to suggest that Stubbe identifies his Arian god with nature, which suggests that Stubbe was a materialist and a pantheist.[40]
His correspondence with Hobbes during the late 1650s points in the same
direction - towards some kind of pantheistic materialism.[41]

   A question immediately arises: if Stubbe thought that Aristotelian philosophy supported false theological doctrine, why would he go to
such lengths to defend Aristotle as a servant of the church? Again the
clue is provided by one of his polemical tracts against Sprat and Glanvill entitled *The plus ultra reduced to a non plus*. There he says that
the primitive church divided into two factions,

> Arians and Catholicks: that the Arians were
> Aristotelians is to me as evident as that
> Mahomet taking the advantage of that faction
> and of the brutal lives and ignorance of the
> Catholicks depending upon the Patriarch of
> Constantinople did advance the sect of Chris-
> tians called Mahometans; and his successors
> the caliphs did wholly employ themselves to

> improve the doctrines of Aristotle and the Peri-
> patetics. So that Aristotelianism, Arianism, and
> Mahometanism issued out of the same parts of the
> world, viz. Alexandria, and the adjacent countries:
> ...[42]

Little enough might be made of this rather incidental comment in which
Aristotle, Arianism and Islam are thrown together in a bewildering intellectual hotch-potch. Little might be made of it, that is, were it
not for the fact that since the early eighteenth century to Stubbe has
been attributed the authorship of a treatise of which various manuscript copies have turned up, entitled *An Account of the Rise and Progress of Mahometanism*. A printed edition of the treatise has been a-
vailable since 1911, but few scholars have troubled to look at it, and
none at all who have dealt with the question of Stubbe's attacks on the
Royal Society.[43] More is the pity of this neglect because the treatise
clarifies the issues. First it can be said with assurance for the
first time that Stubbe was the author of the treatise (I have proved
his authorship elsewhere).[44] Now that we know it is his, it can be
made to throw considerable light on the curious nature of his commitment to Aristotle and the question of his Arianism.

In his treatise Stubbe traces the origins of Islam to the Arianism
of Hellenistic Alexandria. This Arian Christianity, Stubbe argues, was
true primitive Christianity.[45] So Stubbe is claiming something that
would make the hair of the pious stand on end, that the pure Christianity survives, if at all, in contemporary Islam.

What then of orthodox Christianity, the religion asserting
Christ's divinity and the Trinity, where did that come from - or, as
Stubbe might ask disingenuously, where did Christianity go wrong? It
was corrupted from two sources - first an inundation of pagan converts
and second the Emperor Constantine. The converts, being pagan, needed
to be convinced by arguments familiar to their heathen heads, and so in
order to accommodate them Jesus was said to be divine and the Trinity
was invented - all with help from Platonism and a corruption of Aristotle.[46] Thus, as we have seen, Stubbe can argue in Socinian fashion
that orthodox Christianity is false to the extent that it departs from
scripture in the direction of Plato and a debased Aristotelianism.
Constantine and some of his successors did their part too in leading
Christianity astray: they made the church rich and powerful by allowing it to enforce uniformity and to persecute schismatics.[47] Worst of
all, the Catholic clergy claimed independent spiritual authority and

justified it by resort to the notion of an apostolic succession which
was supposed to allow divine authority to be communicated to successive
generations of clergy. Again this communication of spiritual power
rested upon a Platonic and Aristotelian metaphysic of the spirit which
was both unscriptural and fraudulent.[48] Almost certainly Stubbe's understanding in this regard derived from Hobbes and specifically the
fourth part of *Leviathan* where he argues that the clerical claim to
spiritual power is a pretense rooted in Scholastic philosophy.[49] In
this regard Stubbe's Restoration Socinianism was connected to his earlier quasi-Hobbesian Erastian Independency - and the linkage was firmer
than this.

By contrast to the unscriptural and ignorant orthodox Christians
or Catholics, as Stubbe calls them, primitive, scriptural Christianity
survives and flourishes first in Alexandria and adjacent parts and
then, because Mohammed takes it up, it thrives wherever Islam
spreads.[50] Such Christianity is at once Arian and properly Aristotelian.[51] The Arian faith consists of a simple creed based upon belief
in one God and in his justice in dispensing rewards and penalties in
the afterlife exactly appropriate to one's conduct on earth (Stubbe
was not an atheist).[52] The religion is free of abstruse notions like
Christ's divinity and the Trinity that must be shored up by a false and
weighty metaphysic.[53] Beyond this doctrinal minimum one is free to
worship and think as one chooses. There is no separate clerical caste
claiming independent spiritual authority, again justified by a false
metaphysic. To the extent that there is a leader of the church he is
identical with the civil sovereign - a Hobbesian touch.[54] This freedom
from abstruse doctrine, clerical pretension and false metaphysic means
that science and learning can thrive, all under the aegis of Aristotle
and his commentators.[55] Here is the true Aristotelianism, to which
Stubbe subscribes, as distinct from the debased version of the Catholics.

So we are ready to answer our question: Stubbe can defend Aristotle at such lengths as he does, despite the identification of a certain kind of Aristotelianism with false theological doctrine, because
the Aristotelianism to which he is ultimately committed is not that of
the Catholics but is rather the one identified in his mind with Socinianism or Arian belief. As such it chimes with his view of true Christianity - first a civil religion with the secular authority at its head
enforcing a doctrinal minimum, a set of essential beliefs, and allowing

toleration beyond that and, second, a civil religion in which there
would be no separate clergy claiming their own exclusive spiritual authority to preach, teach, discipline and punish.

One might well ask now why did Stubbe take such pains to defend
the established church supported by what he regarded as the debased
Aristotelianism of the Catholics which so appalled him? The answer of
course lies in the context. During the Restoration he was not free
openly to publish his Socinian views; they had to be consigned to a
clandestine manuscript, *The Rise and Progress of Mahometanism*, quietly
circulated among friends.[56] In his public advocacy he had to be cautious, hence his resort to subterfuge. There is another, more interesting reason for his willingness to defend conventional (corrupted)
Aristotelianism: the Catholic threat. That was real too, at least in
men's minds and paranoid fantasies, and he may have been quite sincere
when he argued that Englishmen would have to dispute with Roman Catholics on the latter's terms, which meant knowing the Catholic Aristotle,
if they were to stand up to the popish threat.[57] If Stubbe believed in
anything, it was the force of circumstance. His *Account of the Rise
and Progress of Mahometanism* is a secularized history of the church in
which religion, instead of depending upon divine interventions for its
success or failure, grows out of circumstances.[58] If Stubbe saw religious history in these terms, then he no doubt applied the same understanding to the religious situation in Restoration England. He was a
de facto Anglican just as he was a *de facto* royalist. Circumstances
dictated religious belief, at least in its outward profession, just as
they did political obligation. Against the Catholic threat he can
thus argue for preserving a Catholic Aristotelianism. Should that
threat pass, then perhaps men can be brought around to the true, Socinian Aristotle. This is the view, it seems to me, implicit in his sustained attacks on Glanvill and Sprat.

There is evidence in Glanvill's response to support this conjecture and indeed the whole interpretation of the attacks offered here.
With the assistance of John Beale and Henry Oldenburg, Secretary of the
Royal Society,[59] Glanvill wrote a tract answering Stubbe's attacks in
which he says of Stubbe:

> ...he falls upon His Majesty's institution [the
> Royal Society] out of a pretended concern for
> monarchy and religion. The king, he fancies,
> hath erected a Society that will undermine
> monarchy; and those bishops and divines that

> are embodied in it are managing a design to over-
> throw religion: therefore, M. Stubbe stands up
> in a mighty zeal and defends monarchy against the
> king and religion against the divines, no doubt
> with a purpose to do a mischief to both.[60]

Here Glanvill has penetrated Stubbe's subterfuge, and he goes further
than this. He argues that Stubbe's commitment to monarchy is only pro-
visional and expedient, that should circumstances change, he would re-
vert to advocacy of his earlier republican ideal.[61] Finally, Glanvill
also exposes Stubbe's dedication to "Mahometan Christianity," although
whether Glanvill knew the full extent of Stubbe's Socinianism is unclear.[62] So there was one Stubbe in print and another in private and
in manuscript, a public Stubbe and a clandestine one, and Glanvill,
with help from Beale and Oldenburg, caught something of the connection
between the two in the duplicity lying beneath the surface of Stubbe's
attacks.

From Stubbe's perspective the Royal Society promoted popery by
two means, one of which he spelled out in print and the other he left
to subterfuge and communicated clandestinely. Publicly he declared
that the Society was following Campanella's strategy for subverting the
Protestant Reformation and re-Catholicizing Europe. Cryptically, both
in print and manuscript, he suggested that the natural religion of the
Society conduced to popery in another, more subtle way: to the extent
that the Society subscribed to the traditional Christian mysteries and
used the authority of its philosophy to support them, to that extent it
shared in the superstitious faith of the Catholics and was popish in
tendency.[63] The natural religion of the Society, in other words, was
scarcely superior to the debased Aristotelianism of the Catholics with
respect to dispelling false belief and discerning religious truth.

The only remedy lay in a proper understanding of Aristotle, the
historical Aristotle rather than the corrupt Catholic version. This
proper understanding would support pure Christianity, which was an Arian civil religion shorn of "uncouth and phantastical" doctrines, shorn
too of a separate clergy claiming independent spiritual power.[64] The
recovery of the true philosophy, the real Aristotle, would point the
way to this true Arian faith.

From at least early 1666 and the publication of his cryptic attack
on the royal touch and by implication Christ's divinity, Stubbe had engaged in pro-Socinian polemic. Nor did his attacks on the Royal Society represent a retreat into orthodoxy. On the contrary, they, like his

earlier tract, *The miraculous conformist*, could be read, if read closely enough, as masked statements of his Arian views. Such was the way
the polemic had to be conducted in the 1660s, and Stubbe was an expert.
He upheld Aristotle against the Royal Society and thus could claim to
be defending England, church and monarchy, from popery. But for Stubbe
popery had a double meaning (this was of course a seventeenth-century
commonplace): popery was the Roman Church itself, but it was also superstitious Christian or pseudo-Christian doctrine, which might infect
any communion. To the extent that popery meant the latter, and could
lead to the former, Stubbe's defense of Aristotle represented an attack
on the established church because his Aristotle, the true Aristotle,
was a Socinian. But this could not be made clear in print; it could,
however, be read between the lines. So by defending Aristotle, Stubbe
could conceal his real convictions while not being untrue to them, indeed he could feign orthodoxy while insinuating heresy. It was not a
cover-up but an attempt at cryptic utterance, less subterfuge than dis-
guise.

The polemics of the Restoration debate over the Royal Society were
far more complicated than we have suspected. While some, Meric Casaubon for example, may have defended Aristotle against the challenge of
the new philosophy because it tended to undermine traditional Christianity, Stubbe did nothing of the kind.[65] He attacked the apologists
of the Royal Society because they were wedded to debased and superstitious Christianity and thereby threatened to bring in popery. Stubbe
was clearly not a conservative, as Baker and Jones would have it. Indeed we can now stand their interpretation on its head: from Stubbe's
point of view Sprat and Glanvill were the conservatives for sharing the
superstitious Christianity of the Catholics. He, on the other hand,
turns out to be a radical, and radical in both senses of the term: he
sought to return to what he thought was original - radical - Christianity in order to build the Reformation upon radical, albeit ancient,
foundations.

There is a continuity then between Stubbe's career before and after 1660. His radicalism survived the Restoration. Indeed he is an
example of how the radicalism of the 1650s not only survived but flour-
ished underground and was transformed between 1660 and 1688-89. But
that is a story for another day.[66]

There is one question, however, that we should not postpone. What
does this revisionist approach to Stubbe say to a workshop devoted to
"science, pseudoscience and society"? Baker and Jones depicted Stubbe
as defending Aristotle against the new philosophy whose scientific ad-
vances had rendered Aristotle's science obsolete, as, in other words,
defending pseudoscience against scientific progress.[67] What would
Baker and Jones make of the fact, as it now turns out, that Stubbe defended a backward science for the sake of a radical theology that looks
forward to the pagan naturalism and materialism of the Enlightenment?[68]
From Stubbe's point of view Glanvill and Sprat's natural philosophy and
religion were decadent, and it was he who was in the vanguard. The
picture has suddenly become extremely complicated, and, as a result, it
is no longer possible to see the new science as being unqualifiedly
progressive and Aristotelianism as outmoded in every respect. Baker
and Jones imposed upon the controversy between Stubbe and the Royal Society the categories of science and pseudoscience and missed the historical reality.

As this new understanding of Stubbe's attacks reveals, the Royal
Society had wedded the new philosophy to orthodox Christian theology,
while Stubbe was busy undermining that theology in the direction of
Enlightenment paganism and skepticism by an appeal to Aristotle's authority. In order to see this, I have abandoned the opposition, drawn
by Jones and Baker, between science and pseudoscience and subjected
the evidence to a contextual reading. That is to say, I have observed
the controversy in the light of the contemporary issues - political,
religious and social - that surrounded it and upon which it fed. A social interpretation of science dissolves unhelpful distinctions between
Restoration science and pseudoscience and thus allows us to arrive at a
clearer, more historical understanding. In particular I hope this paper has gone some way towards breaking down the false dichotomy between
science, progress and Enlightenment, on the one hand, and Scholastic
philosophy, superstition and ignorance, on the other. In the right
hands Aristotle could serve radical purposes as well as Bacon, Boyle or
Newton, just as they could be made to serve conservative ones.[69] It
was always a question of the right hands and not of innate properties.

FOOTNOTES

1 Thomas Sprat, *The History of the Royal Society of London* (London, 1667), pp. 22, 57, 63-76, 94, 130-31, 342-77, 400, 408, 426-29 (hereafter Sprat, *History*). For the most recent treatment of the early Royal Society: Michael Hunter, "The Debate Over Science," in *The Restored Monarchy*, ed. J. R. Jones (London, 1979), pp. 176-228.

2 Henry Stubbe, *Legends No Histories* (London, 1670), Preface, p. 2 (unnumbered). Hereafter *Legends*.

3 Harcourt Brown, *Scientific Organizations in Seventeenth-Century France (1620-1680)* (Baltimore, 1934), pp. 256-57; and [Henry Stubbe], *A Reply to a Letter of Dr. Henry More* (Oxford, 1671), p. 59.

4 Joseph Glanvill, *A Praefatory Answer to Mr. Henry Stubbe* (London, 1671), pp. 3-4 (hereafter Glanvill, *Praefatory*).

5 For Stubbe's supporters in London: *Legends*, Preface, pp. 7-8 (unnumbered); and Henry Stubbe, *The Plus Ultra Reduced to a Non Plus*, (London, 1670), p. 135 (hereafter *Plus Ultra Reduced*).

6 *Legends*, Preface, p. 2 (unnumbered); and *Plus Ultra Reduced*, "Premonition to the Reader" and p. 14.

7 *Legends*, Preface, p. 4 (unnumbered); *Plus Ultra Reduced*, pp. 14, 172-73; and Henry Stubbe, *A Censure Upon Certain Passages Contained in the History of the Royal Society*, 2d ed. (Oxford, 1671), pp. 50-53 (hereafter *Censure*).

8 Henry Stubbe, *Campanella Revived* (London, 1670), pp. 3-4 (hereafter *Campanella Revived*).

9 Sprat, *History*, pp. 63-76, 130-31, 343, 400, 408, 426-29.

10 *Campanella Revived*, p. 2.

11 *Ibid.*, p. 3.

12 *Ibid.*, pp. 12-13.

13 Herschel Baker, *The Wars of Truth* (Cambridge, Mass., 1947), p. 365.

14 Richard Foster Jones, *Ancients and Moderns*, 2d ed. (St. Louis, 1961), p. 262.

15 Perez Zagorin, *A History of Political Thought in the English Revolution* (London, 1954), pp. 159-62; and Anthony Wood, *Athenae Oxoniensis*, ed. P. Bliss, 4 vols (London, 1813-1820), iii: 1067-83 (hereafter *AO*).

16 B. L. Add. MS 32553, fos 5, 7, 18; and Francis Thompson, "Lettres de Stubbe à Hobbes," *Archives de Philosophie*, 12 (1936), p. 100.

17 Thomas Hobbes, *Markes of the Absurd Geometry, Rural Language, Scottish Church-Politicks and Barbarisms of John Wallis* (London, 1657).

18 Thomas Hobbes, *Leviathan*, ed. C. B. Macpherson (Harmondsworth, England, 1968), Parts iii and iv; and B. L. Add. MS 32553, fo 12 v.

19 Hobbes, *Leviathan*, ch. 42; and Henry Stubbe, *A Light Shining Out of Darknes* (London, 1659), pp. 6-26 and 82-85.

20 *AO*, iii: 1069.

21 B. L. Harley 3785, fo 111; Edward M. Thompson (ed.), *Correspondence of the Family of Hatton* (Camden Society, n. s., xxii-xxiii, 1878), i: 49; Marjorie Nicolson (ed.), *The Conway Letters* (New Haven, Conn., 1930), p. 274; and J. R. Jacob, *Robert Boyle and the English Revolution* (New York, 1978), pp. 159-76.

22 Marc Bloch, *The Royal Touch*, trans. J. E. Anderson (London, 1973), pp. 208-13. For background: Gerard Reedy, S. J., "Mystical Politics: The Imagery of Charles II's Coronation," in *Studies in Change and Revolution*, ed. Paul Korshin (London, 1972), pp. 19-42; and Carolyn A. Edie, "The Popular Idea of Monarchy on the Eve of the Stuart Restoration," *Huntington Library Quarterly*, 39 (August, 1976), pp. 343-73.

23 Henry Stubbe, *The Miraculous Conformist* (Oxford, 1666), pp. 10-11, 27; and Glanvill, *Praefatory*, pp. 26-27, 60-61.

24 *Legends*, Preface, p. 3 (unnumbered).

25 Gerald M. Straka, *Anglican Reaction to the Revolution of 1688* (Madison, Wisconsin, 1962).

26 Hobbes, *Leviathan*; C. B. Macpherson, *The Political Theory of Possessive Individualism, Hobbes to Locke* (Oxford, 1962); and Quentin Skinner, "The Ideological Context of Hobbes' Political Thought," *The Historical Journal*, 9 (1966), pp. 286-317.

27 Glanvill, *Praefatory*, p. 6.

28 *Ibid.*, p. 63.

29 *Ibid.*, Preface, pp. 6-8 (unnumbered).

30 *Legends*, p. 14.

31 Henry Stubbe, *A Reply unto the Letter Written to Mr. Henry Stubbe in Defense of the History of the Royal Society* (Oxford, 1671), p. 20.

32 Sprat, *History*, p. 355.

33 *Censure*, pp. 45-47.

34 *Ibid.* For the sixteenth-century origins of this theory: Quentin Skinner, *The Foundations of Modern Political Thought*, 2 vols (Cambridge, 1978), ii: 189-359. For the fact that the theory disappears from print during the Restoration: Julian Franklin, *John Locke and the Theory of Sovereignty* (Cambridge, 1978), p. 88.

35 *Plus Ultra Reduced*, p. 13.

36 *Campanella Revived*, p. 13; and *Censure*, pp. 50-52.

37 *Campanella Revived*, p. 12.

38 Sprat, *History*, p. 355; and *Censure*, pp. 50-52.

39 *Censure*, p. 52.

40 The Royal Society of London, Boyle Letters 2, fo 65 v.

41 B. L. Add. MS 32553, fo 12 v.

42 *Plus Ultra Reduced*, pp. 15-16.

43 Henry Stubbe, *An Account of the Rise and Progress of Mahometanism*, ed. Hafiz Mahmud Khan Shairani (London, 1911). Hereafter *RPM*. For a brief introduction to the manuscript: P. M. Holt, *A Seventeenth-Century Defender of Islam. Henry Stubbe and His Book* (London, 1972).

44 J. R. Jacob, "The Authorship of *An Account of the Rise and Progress of Mahometanism*," *Notes & Queries*, N. S., vol. 26 (Feb., 1979), pp. 10-11.

45 *RPM*, pp. 8-58.

46 *Ibid.*, pp. 28-29.

47 *Ibid.*, pp. 35-36.

48 *Plus Ultra Reduced*, p. 16; and *RPM*, pp. 15-24, 29-37.

49 Hobbes, *Leviathan*, chs 46 and 47.

50 *RPM*, chs iv-v and pp. 145-46, 164-70, 178, 188-89.

51 *Ibid.*; and *Plus Ultra Reduced*, pp. 15-16.

52 *RPM*, pp. 164-68.

53 *Ibid.*, pp. 15-16, 28-29, 164-66.

54 *Ibid.*, p. 178.

55 *Ibid.*, pp. 43-48, 53-57; and *Plus Ultra Reduced*, pp. 15-16.

56 J. R. Jacob, "Civil Religion and Radical Politics: Stubbe to Blount," presented at the Annual Meeting of the American Historical Association in San Francisco, 30 December 1978.

57 For example: *Campanella Revived*, p. 13; and *Censure*, pp. 50-53.

58 For Stubbe's most explicit statement of this view: *RPM*, p. 49. (The whole work is grounded on this assumption.)

59 For the collaboration: Christ Church Library, Oxford, Evelyn MSS, John Beale to John Evelyn, 11 June 1670, and 3 August 1671.

60 Glanvill, *Praefatory*, Preface, pp. 3-4 (unnumbered).

61 *Ibid.*, Preface, pp. 6-8 (unnumbered).

62 *Ibid.*, pp. 48-49, 62-63.

63 *Plus Ultra Reduced*, pp. 11-12.

64 *Ibid.*, pp. 15-16; *Censure*, p. 29; and *RPM*, *passim*.

65 For Casaubon: Michael R. G. Spiller, "Conservative Opinion and the New Science, 1630-1680: With Special Reference to the Life and Works of Meric Casaubon," B. Litt. Thesis, Oxford University, 1968. I owe this reference to Dr. Michael Hunter.

66 Reference 56 above; and J. R. Jacob, *Civil Religion and Radical Politics: Stubbe to Toland* (in preparation).

67 Baker, *The Wars of Truth*, pp. 362-65; and Jones, *Ancients and Moderns*, pp. 244-63.

68 J. R. Jacob, "Boyle's Atomism and the Restoration Assault on Pagan Naturalism," *Social Studies of Science*, 8 (May, 1978), pp. 211-33.

69 J. R. Jacob and M. C. Jacob, "The Anglican Origins of Modern Science (The Metaphysical Foundations of the Whig Constitution)," *Isis* (forthcoming in June, 1980).

DEPLOYING 'PSEUDOSCIENCE': THEN AND NOW

(i)

I want in this paper to extend the discussion that has taken place on the role of pseudoscience in the seventeenth century in an attempt to formulate a general thesis about the role of pseudoscience in modern society. By 'role' I do not mean the significance of a particular pseudoscience for those involved with it, though I will be concerned with this aspect when I turn to the historical focus of this paper, the classic Victorian pseudoscience of phrenology. By 'role' I mean here simply what function the appellation 'pseudoscience' serves for those who deploy the term. And the thesis I wish to forward is simply this: that at least since the beginning of the consolidation and ossification of the capitalist order in the seventeenth century, the label 'pseudoscience' has played an ideologically conservative and morally prescriptive social role in the interests of that order. Further, I want to argue that whenever and wherever we encounter the deployment of the label 'pseudoscience' we are encountering a process of conserving social interests.

This thesis will not, I suspect, raise any eyebrows among historians of seventeenth-century science and society. As Everett Mendelsohn has recently summarized, the driving underground of hermeticism, alchemy and magic in the seventeenth century was very much a part of a process of rationalizing and legitimating the newly altered social order and its intellectual controls. The "socially imposed and self-consciously accepted" distinction between science and pseudoscience was one of the chief means of obfuscating the social and intellectual origins of the new science and its methodology.[1] Thereby the new system of knowing nature was able to establish itself as a value-transcendent touchstone of truth, reason and rationality for the evaluation of all subsequent social, moral and intellectual productions.[2]

If there is any shock value in the thesis I'm proposing it will be for those who see as 'radical' the attacks on, say, Arthur Jensen's hypothesis on heredity or Hans Eysenck's on race and intelligence. The idea that those who paste the label 'pseudoscience' on such activity should be reckoned as conservatives may come as some surprise. The

essay entitled "Scientific Racism and Ideology: the IQ Racket from
Galton to Jensen", written by the self-styled Marxist Steven Rose,[3]
will hardly seem to many (least of all to Rose) an apology for the status quo. I'm also fairly certain that the socialist Monthly Review
Press would not be flattered to hear 'conservative' applied to their
recent publication of Jeffrey Blum's *Pseudoscience and Mental Ability:
the Origins and Fallacies of the IQ Controversy*. So before beginning
to relate my thesis to a specific historical context, I'd like to spend
a moment illustrating how I see these deployments of pseudoscience as
conservative. As will become clear, my intention is far from being to
apologize for non-socialist writings on pseudoscience. Rather, I wish
to reveal the conservative tendency of those labelling pseudoscience
within contexts where that tendency is *least* apparent. In the course
of this exposure I'll also be laying bare the theoretic that informs my
own historical project. Finally, because insightful comments have already been made on Jensen and his critics,[4] and because an incisive
critique of Steven Rose's essay has recently appeared in print[5] - both
of which critiques have greatly informed my own thinking - I'm going to
rotate my comments around Blum's book which has not yet (so far as I
know) come up for critical review.

For starters, let's begin with Blum's definition of pseudoscience
which, for the most part, is characteristic of the way the term is deployed by today's liberal-minded. Pseudoscience, he says, is the
"process of false persuasion by scientific pretense". Its existence
is dependent on the joint occurrence of (1) grossly inadequate attempts
at verification and (2) the successful dissemination of the unwarranted
conclusions drawn from those attempts at verification. "The label
'pseudoscience'", he continues,

> becomes pertinent when the bias displayed by scien-
> tists reaches such extraordinary proportions that
> their relentless pursuit of verification leads them
> to commit major errors of reasoning. Simple tauto-
> logical fallacies, a refusal to consider obvious
> alternative explanations, and the deliberate or
> unconscious falsification of data constitute some
> of the basic features of pseudoscience.[6]

Blum also tells us how pseudosciences work: usually they are associated

> with individual cranks working in isolation to
> promote their eccentric theories. Most instances
> of pseudoscience conform to this description and
> involve nothing more than the development of an

> idiosyncratic cult. When social circumstances
> are favorable, however, the perpetrators may
> acquire sufficient resources and respectability
> that they appear to the society generally as
> legitimate scientists rather than cranks.[7]

The obvious historical inadequacies of this use of pseudoscience
scarcely need dwelling upon. It is difficult to regard Newton, for
example, as a crank because of his study of alchemy; just, as Barry
Barnes has demonstrated, it is difficult to regard Jensen's scientific
work as any less a product of a disinterested search after truth than
that of Jensen's scientific opponents W. Bodmer and L. Cavalli-Sforza.[8]
As for the isolation of the 'crank', while it is hardly the case, at
least before the second half of the nineteenth century, that anyone in
scientific discovery worked in close professional association,[9] on the
other hand it can no more be said of the 'crank' than of the orthodox
scientist that s/he operates in a social or cultural vacuum. Nor do I
think that with this example of the use of the label 'pseudoscience' we
need to deliberate any further on its consistently pejorative nature.
It is rather to the unstated ideological implications of Blum's deployment of the label that I would like to turn. This is not because the
term 'ideology' is absent from the book; on the contrary, it is precisely because Blum discusses at length what he calls 'the ideological
dimension'[10] that he invites discussion on the nature of ideology in
science.

The debate on this issue is by now familiar and has even acquired
a certain sterility. The one thing that does seem to have been largely
overcome in the wake of these debates, however, is the former difficulty of agreeing on a definition of ideology. Historians of science
of many different political persuasions would now generally agree that
by 'ideology' we mean worldview, or the partial view of nature and human nature expressed by a group or class which informs perception and
conceptualization. It is not meant that ideology must be false or
that the reality of nature is necessarily distorted through ideology.
The usage is the marxist-informed but non-pejorative, non-epiphenomenal
one denoting the social, political, metaphysical, and theological or
philosophical superstructure that must accompany every economic system.[11] According to this view, science as an intellectual formation
and ideology as worldview can never be separate realities or autonomous
'things' merely interacting, but must always be mutually constitutive
of each other or interpenetrating to form a seamless web.

For Blum however, in common with most other writers on the IQ controversy, ideology implies distortion and falsity. It is the external
social and political thing that lurks in the shadows and infiltrates
objectivity; it is what can render science into pseudoscience and
'scientism'.[12] Not surprisingly, the two stock examples of scientism
in modern Western literature, Social Darwinism and Lysenkoism, find
extended discussion in Blum's study and as elsewhere these are exploited as illustrations of how science truncated and simplified may be
falsely pressed into the social and political domain - that is, into
the domain of 'ideology'.[13] Such a narrow trivial use of the term
'ideology' (specifically 'meritocratic ideology'[14]) deflects attention
from the real ideological function of Blum's exposé of pseudoscience,
for it conceals what is implicit in it: the belief that there actually
is out there somewhere a real, objective (non-ideological) body of
truth called Science. By dogmatically asserting the existence of an
imagined opposite, - truncated, 'ideological' pseudoscience - Blum
judges as he names and obscures the fact that modern science itself was
and remains the product of men in the activity of historically specific
social relations. As Lukács pointed out as long ago as 1923, modern
science and its methodology were the chief instruments in reifying (or
making to appear as objective) the social relations upon which our
science was reared.[15] Through the development of science the power relationships of capitalist society were concealed and man was forced to
submit to the dictates of what would then appear as the impersonal laws
of nature. In other words (to use the terms of the labellers of pseudo-
science), the methods of science were 'misapplied' in the first place.
Persons such as Blum in forgetting this and in forgetting that the reification of reality into science/pseudoscience was the active creation
of bourgeois or 'proto-bourgeois' intellectuals, thus conduct the whole
of their argument on the terrain of this reified reality. Their arguments can therefore only further legitimate the tyranny of the 'objec-
tified' concepts that sustain capitalist social relations.[16]

In effect Blum merely extends the long tradition of left-wing positivism which, far from seeking to alter the fact/value, truth/false-
hood, use/abuse, science/ideology dichotomies, actually exploits them
in the hope of altering values. While this, it seems to me, is excus-
able where we find it among early nineteenth-century radicals caught in
the snare of progress and without a coherent theoretical understanding
of their situation, it is inexcusable today in light of the persistent

concern with *praxis* among socialist historians of science - the concern,
that is, with the need to keep the intellectual activity of socialist
writing unseparated from socialist theory. Eschewing marxist *praxis*,
Blum unwittingly undermines his honorable anti-racist and socialist intentions by Whiggishly (in a negative sense) making those intentions
the reason for historical investigation. As so often is the case, good
intentions on their own blind the intellect, so that instead of contributing to the solution of perceived iniquities the problem is only com-
pounded. Blum's exposé of pseudoscience only subserves the social organization that sustains the problem against which his humanistic ar-
rows are so passionately directed.

Moreover, because of Blum's socialist teleology he is unable to
enter into any kind of 'relativist' perspective on scientific knowledge,
such as that provided by enthnomethodologists of science.[17] Completely
foreign to Blum is the latter's Wittgensteinian interest in showing how
scientific meaning and conceptualization are dependent upon the social
context (and political interests) in which they are embedded. Unlike
the ethnomethodologists, Blum, in taking 'correct science' as the sub-
jet to explain 'false science' is committed to a sociology of error
rather than to a sociology of knowledge in which scientific truth is to
be seen as a cultural artifact socially negotiated and organized in a
specific milieu.

(ii)

Rather than falling into the same trap I want, in turning now to
the history of phrenology, to delineate an actual historical process in
its own context. Like Blum, I too have a cautionary motive and dissent
strongly from historical agnosticism. But unlike Blum I perceive that
my cautionary tale on the deployment of the term pseudoscience can be
achieved only by treating the historical facts of phrenology in a non-
evaluative, non-teleological, 'relativist' and negotiational way.
Since nearly everything that has been written on phrenology has not
taken this approach, I hope it will enhance the instructive value of
this paper if I bounce off the history of phrenology from its received
historiography.

Until very recently most historical accounts of phrenology were
written in the positivist use/abuse framework.[18] The aim, conscious or
unconscious, was not the writing of history, but the legitimation of

the wisdom of value-free modern science. Consequently these accounts have given scientistic capital to dozens of other attacks on perceived scientifically-related bogies. Karl Dallenbach, for instance, in an article in the *American Journal of Psychology* in 1955 sought to deride the scientific pretentions of psychoanalysis by likening it to phrenology, and the same path has more recently been trod by the Nobel scientist turned literator P. B. Medawar.[19] That a similar use of phrenology is overlooked by Blum is, I suspect, only a measure of his historical ignorance. His better-informed alter ego, the neurobiologist Steven Rose does not miss the opportunity. Twice in his essay on the IQ racket he refers to Eysenck as operating in "the manner of a nineteenth-century phrenologist" as if this comparison utterly vanquished Eysenck's credibility.[20] In fact it invalidates nothing, since Eysenck himself in his popular works on the 'facts and fictions in psychology' justifies his scientificity by a similar contrast with phrenology.[21]

The positivist motives of both the left and the right in so deploying phrenology are patently obvious, and it has been partly because of this that revisionist historians of society and medicine have felt a strong need to rescue the subject.[22] These historians have performed a good service: they have shown (much in the manner of Darnton's study of mesmerism in France)[23] that phrenology in its social context was far from being the outlandish nonsense that Whig historiography implied. By observing the interest in phrenology of countless illustrious contemporaries from A. R. Wallace, Robert Chambers and Samuel Smiles, to Comte, Spencer and Marx, they have shown that phrenology's impact and significance at both popular and intellectual levels was far deeper than had been previously assumed.[24]

Yet, paradoxical as it may seem, it cannot be simply inferred from this that phrenology was therefore scientifically legitimate in its context. Since the contextual evidence for phrenology being regarded in Britain as "a sick man's dream"[25] is in fact overwhelming, the Whig historians of phrenology were perfectly justified in calling phrenology pseudoscientific. From the moment of its first appearance in Britain, Gall's theory of brain was regarded as physiologically, craniologically and philosophically visionary and dangerous. Thomas Brown, in a well-known review of the theory in the *Edinburgh Review* of 1803, set a precedent for many far more passionate commentators with his conclusion that it was an unconvincing "species of phsyiognomy" unjustified by anatomy. Before 1820 phrenology was regarded with disdain in nearly every medical and literary journal while in popular prose and poetry, theatricals and cartoons it was satirically lambasted. And this was far from being just an initial Anglophile reaction to a Franco-Germanic theory. A sustained stream of irreproachable evidence rationally deducing phrenology's scientific failings flowed from Britain's most distinguished contemporary scientists and philosophers. Though nationalism was not absent from these critiques, it was not essential to them and was not in fact much relied upon. More typical was the dispassionate critique such as that presented by the physiologist Peter Mark Roget in his article on "Craniology" written in 1818 and published in the sixth edition (1824) of the *Encyclopedia Britannica*. In this, the case against phrenology was divested of emotional anti-materialism and anti-fatalism and was conducted on the basis of supposedly sound, neutral physiological evidence. Gall's basic physiological premises and his cerebral anatomy were even praised, while his theory of cerebral localization and the attendant craniology were torn to shreds.

By 1842, when Roget's article was republished unaltered in the seventh edition of the *Britannica* under the new title of "Phrenology", the scientific and scholarly backing for its opinions had been impressively expanded. In books, in articles in leading scientific and literary journals, dictionaries and encyclopedias, and in lectures delivered in Royal Societies, the doctrine was attacked in the 1820 and '30s by such figures as Dugald Stewart, Charles Bell, Sir Everard Home, Francis Jeffrey, Sir William Hamilton, James Cowles Pritchard, John Bostock and James Copland among many others. And, as if these weren't enough, at the same time came the well-substantiated opposition to the doctrine by the French physiologists François Magendie and J. P. M. Flourens, the Berlin physiologist Karl Rudolphi, and the professor of physiology at Columbia University, Thomas Sewall. Whig historians, then, however misguided in their motives, were thus quite correct to call phrenology 'pseudoscientific' in the sense of it being 'discredited knowledge'. Their 'only' mistake was to suppose that because of this phrenology must be unworthy of serious historical study.

Just how wrong this supposition is becomes manifest as we swing round to what the revisionists assay, although it should be added that the revisionists neither understand nor appreciate the implications of the Whigs' error. Their primary intention after all is with the writing of correct history, not with the lessons nor the sociological insights to be derived from it. This aside, what they *have* amply proven is that while the highly qualified opposition to phrenology was being mounted, the phrenology movement was advancing by leaps and bounds. There is not time here to spell out the many facets of this movement nor is it so necessary given the many recent publications. Suffice to say that the institutionalization of phrenology was significant both in the narrow sense of formal organizations at national and local levels and (as I will amplify below) in the sociological sense of establishing a knowledge/behaviour base for standardized collective reference and action.[26]

An important point that I would like to draw your attention to, is the fact that the intellectual leadership of this movement was heavily weighted with medical men and that it was they who were most influential in popularly disseminating phrenology among the self-styled 'thinking classes' between the 1820s and 1840s. One finds, for example, that in the national Phrenological Association (which was formed belatedly in 1838) of the 218 members for whom we have occupations, the largest proportion - 33% - were medical men. Local phrenological societies reflect a similar or great preponderance of this professional group, and according to one phrenological statistician, at least a third of the one hundred writers on phrenology in Britain that could be identified in 1838 were either physicians or surgeons.[27] The most notable of these medical proponents of phrenology were William Lawrence, surgeon to and lecturer at St. Bartholomew's Hospital, John Elliotson, physician and lecturer at St. Thomas's and later Professor of Medicine at London University, and Thomas Wakley, the editor of the *Lancet*. Leaving aside the influence of such phrenological alientists as Sir William Ellis, John Conolly and W. A. F. Browne, who pioneered the non-restraint moral therapy in Britain,[28] and considering only what Wakley accomplished for phrenology in the *Lancet* (where over 600 pages are devoted to science between 1824 and 1851), and what Lawrence and above all Elliotson and his disciples accomplished through their textbooks and lectures in the major London medical schools and in the new provincial medical schools, we can readily understand how difficult it was for the rising generation of medical students to escape contact with the new theory of brain and science of character. Moreover, when we read these men's rationales for phrenology in the light of the virtual absence of any medically-useful theories in mind, there is little difficulty in explaining in terms specific to medical science why the profession was so widely responsive to phrenology in the early nineteenth century.[29]

I do not put this seemingly intellectualist explanation forward as
the full reason for the medical profession's support of phrenology; in
fact the explanation is not really intellectualist at all, but compel-
lingly social, since phrenology maximized the competences of medical
men in predicting mental phenomena. Here, however, I merely want to
underline that in the context the reasons for supporting phrenology
could be seen as entirely 'rational' or scientific. Noting this and
noting the preponderance of medical men in the movement should forestall the characterization of the phrenology movement and its scientific opposition as akin to the misconstrued depiction of science during
the French Revolution, that is, with capital 'S' genuine Science on the
one side and a crude popular scientism intimidating that 'real' Science
on the other.[30] It is certainly true that during the Industrial Revolution of the early nineteenth century there was an impact of scienti-
fically-mediated 'styles of thought' which was as profound - if not as
violent - as the collision between the so-called 'Newtonian' and 'anti-
Newtonian' cosmologies during the French Revolution.[31] However, there
is no advantage in transcending the historical context and branding one
of these styles of thought (that of the antiphrenologists) 'scientific'
and the other (that of the phrenologists) scientistic. As is indeed
suggested by the fact that it was Britain's most distinguished scientists and philosophers who conducted the attack on phrenology (and
mightily sustained that attack) the antiphrenologists no less than the
phrenologists can be seen as having been anxious to defend interests
above and beyond what we would normally call scientific.

What these interests were becomes clear when we socially map the
participants.[32] In brief (and comparing here only the scientific-
medical opponents and supporters of phrenology) what stands out in bold
relief are the differences in age, economic and social security, religious affiliation and politicization. The supporters of phrenology
were on the average twenty years younger than their opponents. They
were socially marginal and economically insecure: few had upper middle
class or aristocratic backgrounds, had had elitist educations or belonged to elitist institutions, as did many of the antiphrenologists;
and while some managed to struggle into university positions during
the period in which they advocated phrenology, these were few in number and were generally confined to the less prestigious postings. In
terms of religion: as opposed to the antiphrenologists' High Church
involvement and firm attachment to socially undisturbing Paleyian

natural theology, virtually all of the supporters of phrenology were
either avowed materialists, proponents of 'the grander view' of the
Creator (sometimes bordering on token deism), or were vocal dissenters.
These positions on religion were at the same time political postures
for which party labels serve as insufficient guides. For the most part
the supporters of phrenology can be seen as Radical partizans, while
the antiphrenologists can be seen as having a more conservative attachment to Whig reformism.

The fuzzyness of this last distinction reminds us, however, that
the social and ideological conflict between phrenologists and antiphrenologists cannot be reduced simply to one of class nor exclusive cosmologies and worldviews. It is for this reason that I have deliberately
relied on Mannheim's protean term 'styles of thought'.[33] To refer to
exclusive cosmologies might imply that the phrenologists were providing
the intellectual rationales for social structures and relations totally
different from those of the antiphrenologists. But this was not quite
the case, even though between the two groups elements of a genuine conflict between aristocratic and bourgeois interests can on occasions be
isolated. The antiphrenologists for the most part shared with the
phrenologists an optimistic uniformitarian faith in gradual progress,
in social homogeneity, stability, and hierarchical order, and shared
with the phrenologists a non-literalism in Biblical matters and a view
of a lawful rational universe presided over by a benign Deity.[34] Thus,
although the reaction to phrenology by the orthodox scientific-philoso-
phical community (especially in Edinburgh) bears a striking resemblance
to the reception given to Velikovsky's ideas by our own scientific community,[35] in actuality the outlook of phrenologists and antiphrenologists was very far from being parallel to that of today's catastro-
phists and uniformitarians. Though the conflict involved metaphysical
and metaphorical sound and fury, it is an over-simplification to say
that worldviews were in collision. The history of the conflict is more
subtle than that, involving in part the rhetorical build-up of caricatures which were then used by marginal men offensively and defensively
for the rationalizations of the industrial secular form of bourgeois
capitalism. These caricatures distorted and polarized the views of
phrenologists and antiphrenologists making the phrenologists appear as
progressives for an open society allied to the mobile fortunes of com-
merce, and the antiphrenologists appear as the retainers of a static
and privileged aristocratic society allied to inherited agrarian wealth.

What in most cases were differences between different factions of the
same class over *degrees* of social re-ordering and the *pace* of social
change were, through the dialectical spiral of the phrenology debates
in the context of the further emergence of urban industrial society,
construed as competing class views over *kinds* of social orders. The
point is historically important but is not of central concern to us
here, for caricature or not, the consequence of the debates over phrenology was the same: the impossibility for any participant to remain
politically, philosophically, or theologically detached. No matter
whether the advocacy or denunciation was polemical or factually dispassionate, it could never be intellectually autonomous or be *just*
scientific and medical.[36]

Once we disengage phrenology from the clutches of the use/abuse
model and perceive it in this way as a competing style of thought or
social and ideological strategy, we can identify the conservativeness
in the contemporary labelling of it as 'pseudo'. The labelling can be
called 'conservative' not only because at a political level the phrenologists happened to appear as more radical, but also and more fundamentally, because those in or identifying with the scientific and philosophical elite who deployed the label were seeking to discredit what they
saw to varying degrees as socially threatening and, by this process,
legitimate and defend certain human and social interests and the neutrality of the science that served them. It is the basic defence of
interests, rather than the specific political colour of those interests
that warrants the use of the term 'conservative'.

(iii)

It is not, however, for this trivial illustration of my thesis
that I have led you thus far. I have taken the course I have in order
that I may now turn to the far more interesting and - at the political
level at least - seemingly paradoxical recapitulation of the thesis.
What I now wish to reveal is that the phrenologists themselves were involved in labelling pseudoscience and that this labelling also acted
conservatively both in the obvious sense of a defence of social interests and in an historically specific political sense.

On the first point - that phrenologists were involved in labelling
pseudoscience - we have explicit confirmation from as early as 1821
when an Edinburgh phrenologist likened to alchemy the methods of the

elitist mental philosophers.[37] The reigning mental philosophy, it was
claimed, proved itself to be antiscientific since it was wanting in
both empiricism and utility. Phrenology was juxtaposed to this, reflecting the image of true science through its faithfulness to the
methods of Bacon. The superficial means by which phrenology allowed
the mysteries of man's nature to be empirically observed (literally
from the surface of the skull) highlighted the difference with as it
challenged the elite's mysterious metaphysical introspections on man.
And while the fruits of the phrenologists' knowledge, which to them
proved its truth, could be seen as directly applicable to the practical
affairs of men - indeed, could be shown to be useful in ameliorating
the existing condition of society - the obscure knowledge of the elite
was useful only to the elite, thereby proving its subjectivity. In
the manner of seventeenth century exponents of Baconism, phrenologists
argued that real science or real truth as derived from nature was open
to the observation and common sense of all. Pseudoscience, on the
other hand, they said (in the manner of Blum), was the production of
closed associations and obscure practices. The fact that antiphrenologists had to resort to cruel vivisection in their attempts to disprove
phrenology made plain, among other things, the essential immorality of
the methods of pseudoscience.[38] By virtue of the wide public appeal of
such rhetoric, the phrenological defence of the image of science was,
as Steven Shapin has made clear,[39] inseparably a defence of the rights
of 'the people' against the rule of privileged groups. Thus in attacking the philosophic basis of the elite's science both the rule and the
legitimation of the rule of the socially privileged was discredited.[40]

Turning now to the second point - that these phrenological attempts at labelling pseudoscience were both socially and politically
conservative - this will not (at the political level at least) be any
more apparent than my earlier and similar claim against the labelling
of pseudoscience by the socialist Blum. This can only be seen in fact
if we take proper advantage of historical hindsight. To remain fixed
in the limited historical context in which phrenologists made their
deployment of pseudoscience is much the same as taking at face value
Blum's attack on pseudoscience and so failing to escape the uncritical
context provided by him. As the latter would lead us to see Blum as
'radical' so, if we do not look beyond the context of pre-reformed
Britain, we can only see the phrenologists as progressive liberals. I
have already noted that the image of phrenologists as doing unilateral

battle with the forces of aristocracy is in large part the distortion
created through the phrenologists' own rhetoric. But even to the extent that this was a true picture of reality, it obscures the fact that
the phrenologists were not innovating so much as reaffirming the social
relations of capitalist wage-labour that had been established in the
seventeenth century.[41] This is signified through the phrenologists'
deification of Bacon and the whole congruence of their rhetoric with
that of the upwardly mobile in the seventeenth century. And it is made
manifest upon an examination of the structures and functions of the
brain that phrenologists popularized - popularized with particular zeal
among refractory workers. This was a brain of quantifiable mental
faculties, a body machine that ideally functioned like a well-managed
factory whose maximum productive capacity (i.e. maximum success in life)
was contingent upon an understanding of and compliance with its rigid
division of labour and hierarchical organization. In effect, to believe
in phrenology, was to internalize the authorities and disciplines of
the capitalist labour process and to see as natural the social relations that went with it. In saying that through this popularization
phrenologists only reaffirmed pre-existing social relations, I am not
suggesting that we under-estimate what was *new* in the nineteenth century - urban industrialization; only that we do not submit to believing
that the new industrialization created an entirely new economic basis
for social relations. In reality industrialization was only a new cata-
lyst for a reassertion, especially by petty-bourgeois groupings, of
pre-existing social relations - albeit in a different industrial context. It is in this sense that the phrenologists' labelling of pseudo-
science was politically conservative, though their immediate context
lends to them the appearance of being radical counter-culturalists.[42]

Naturally, when we move into the period in which the ideology of
industrial bourgeois capitalism was becoming the ruling ideology (after
the settlement of 1832), we would expect that the conservative role of
the phrenologists' labelling of pseudoscience would be more apparent.
This is exactly the case, but that it should have been possible at all
to use phrenology in this way necessitates further confirming our thesis by enquiring into the science's career in the later period.

By the late 1830s phrenologists were beginning to find that their
'outsider' status had considerably altered. Gradually they themselves
were beginning to command power, status and wealth. This was not of
course a simple matter of co-option. The social order had been

partially reformed in response to industrialization and to the pres-
sures of the meritocratic ideology for which the phrenologists served
as important functionaries.[43]   The antimeritocratic ideology, symbol-
ized for the phrenologists in the arrangements and productions of
Hanoverian science, was beginning to seem weaker. Talent and ability
it now appeared were beginning to be appreciated more than wealth and
inherited station.

It was in this context of an expanding middle-class and encroach-
ing meritocratic outlook that the major intellectual organs began to
shift ground in their opinions of phrenology.[44]   "Inasmuch, as at this
moment, the phrenologists appear to be the most strenuous advocates for
putting society on an improved footing," the liberal-Benthamite newspaper the *Examiner* stated as early as October 1831,

> we must say that all haters of the present unequal
> and unnatural distribution of power and knowledge
> must feel indebted to them for their exertion...
> ...if phrenology conduces to the formation or
> spread of such opinions as it maintains, that phre-
> nology strictly cannot be a bad thing.[45]

Over the next decade the opinion that phrenology was a 'good thing' socially, could be taken from innumerable journals, many of whom were the
same who had earlier expressed contempt. The success with which phrenology was extending bourgeois hegemony over the working class was especially gratifying to all manner of bourgeois publicists. This height-
ened social appreciation of phrenology allowed phrenologists in turn to
refer to the flight of former bigotry and prejudice, though they often
mistook what was appreciated for what it culturally exemplified as an
appreciation of the scientific truth of phrenology. Nevertheless, it
was difficult for those coming to value the social worth of phrenology,
to make that appreciation co-exist with a continued disdain for phrenology as a science. Typically therefore one finds contemporaries ac-
quiescing to "what we [now] conceive to be the more rational and moderate phrenological views" even to the point of admitting "to a certain
extent...cranial characteristics".[46]

But this acquiescence was not simply the result of the altered social structure; it was also a measure of how well the phrenologists had
promoted their science as reflecting not just the true image of science
but the socially respectable image of science as well. With the characteristic ambiguity of the upwardly mobile, phrenologists sought to
denounce the so-called anti-science of the elite while at the same time
create for themselves the appearance of elitist respectability. For

this it was necessary to make their own knowledge conform to the polite
and value-free image of science held by respectable society - a need
that became all the more pressing as the social values being advanced
through phrenology moved to the normative center. If the phrenological
expression of human nature was to effectively serve bourgeois hegemony,
it would have to gain the appearance of being more scientific than ideological; its own social interests, in other words, would have to be
better disguised. Thus, as the social structure was being actively
shifted by reformers (many of whom were involved with phrenology), so
too the image of phrenology's scientificity was being further worked
upon.

I have already mentioned that medical spokesmen played the most
important role in phrenology's popular dissemination. This was because,
in spite of their rejection of elitist forms of knowledge, they could
still lend to phrenology an air of scientific integrity and 'natural-
ized' authority.[47] It has to be remembered that to the majority of lay
persons the antiphrenological literature written by the scientific
elite was not readily accessible. Far more available and provoking
were the bigoted religiously-motivated attacks. Hence to those seeking
to establish society on a more 'rational' footing, the very presence of
members of the medical profession could be reassuring. But it was the
supposed value-neutrality, of the medical knowledge, that gave the
greatest leverage. Since the leading opponents had also to rely on
medical knowledge, it was not in the capacity of the opposition to reveal all medical knowledge as socially mediate. The sanctity of medical science remained inviolate since it was in everyone's interest
that it should appear so. Thus journals such as the *Gentleman's Magazine*, which began in the 1820s to cater to the new breed of gentlemen
in the Literary and Philosophic societies and such like, could simply
draw on the expertise of the medical profession to reverse their earlier hostility and justify their new phrenological recommendations.[48]

In the lecture halls the scientific expertise and supposed objectivity was visibly reinforced by the very artifacts of medical science -
the skulls, busts, casts, and charts of brains with which lecturers
surrounded themselves. Dr. Spurzheim (Gall's former collaborator) who
not only brought the science to Britain but did more than any other
itinerant to disseminate the science in the 'Lit & Phils' in the 1820s
was particularly fond of so equipping himself and, as critics in the
scientific elite pointed out, "few were prepared, by previous knowledge,

to estimate the real merits of his demonstrations".[49] Outlines of lecture courses on phrenology in the 1820s in literary and philosophical
societies indicate very clearly how the case for phrenology appeared to
be argued by facts alone and how upon these facts the concluding highly practical remarks seemed sensible and convincing.[50] It is also
worth mentioning at this point that the central factual illustration in
all the phrenological literature, the phrenological head itself, had
become scientized in the 1820s. In Gall and Spurzheim's early diagrams
the head was portrayed in a natural lumpy manner and the countenance
was of a real person. In the 1820s however stylization took over: the
head now came to be unrealistically depicted as spherical and counte-
nanceless and altogether reminiscent of an illustration from mechanics.
This dehumanized abstraction of a head, seeming to symbolize the perfect, the controlled, and the absolute became, like the unexpected sim-
plicity of $E=mc^2$, a magic formula for making a new view of reality sacrosanct.[51] Like other objects made sacred, the phrenological head
through this very act of scientificity had the humanness of its creation quite literally smoothed away. It became - or at least the intention was there to make it become - a reified object, mystifying the
ideological origins of its construction. And let us not overlook the
significance of the fact that in a period when the division between
mental and manual labour was entering its modern form, that it was the
head and *not* the hand or the heart or the genitals that was being made
the sacred object of a movement so closely linked to the advancement of
the industrial capitalist order.[52] What better means to justify the
arrogant superiority of intellectuals over the mass of exploited 'hands'
than to elevate and celebrate the head in the body hierarchy? Of the
many levels at which phrenology can be seen as mediating the emergent
social relations of industrial capitalism, none is so opaque yet so
open to transparency as the phrenological head itself.[53]

     But to return to the more conscious, rhetorical level of the at-
tempt to obfuscate the ideological origins of phrenology, I want to
draw your attention, lastly, to the phrenologists' attitude toward
charlatanry and demagogery. As will be readily appreciated, the success of phrenology as an elitist scientific movement intelligently in-
forming social opinion through 'value-neutral' knowledge, could hardly
be assisted either by the presence of apolitical phrenological mounte-
banks out for financial gain or by political militants exploiting phrenology in the interests of atheism and socialism. With respect to the

charlatans: even though phrenologists practiced little else at their
meetings than craniology, based all of their practical remarks on this
reading of heads and legitimated all that was unique in their social
reform program on cranioscopic diagnosis; even though their attack on
the arcane, introspective and mysterious methodology and productions
of the old elite was mounted from the perspective of phrenology's easy
access to reality; and even though phrenologists prided themselves on
the "astonishing accuracy" with which they could read heads and told
their audiences to look to their own heads for the real proofs of the
science (and of course gained their most zealous converts in this way);
they had, *pari passu*, to convey to the respectable public from whom
they sought recognition the impression that phrenology was not a superficial means of character reading but was in fact a sophisticated
branch of human physiology requiring deep study.[54] With respect to the
demagogues: even though phrenologists themselves argued militantly
against the elite in the name of 'the people', had among their ranks
many committed atheists and materialists and even some sympathizers
with Owenite socialism, and even though they argued that the great
merit of phrenology was its universal application to human affairs,
they had simultaneously to convince their respectable audience that
radical political and social uses of phrenology were really scientistic
'misapplications' of a science that in and of itself was neutral. For
the perpetration of this image the vulgar phrenological fortune tellers
and the socially and politically threatening radical atheists and so-
cialists were in fact a great convenience, in much the same way that
the lower-class radical activity and interest in 'marginal science' in
the seventeenth century that had assisted the bourgeois revolutionary
cause became, in turn, the convenient source for the legitimation of
the bourgeois social and scientific hegemony. Phrenologists, by themselves deriding 'bumpology' and 'phrenological palmistry' as corruptions of their science and by attacking the infidel and socialist uses
of phrenology as 'misapplications', restaged for industrial society the
prestidigitation of seventeenth-century scientific rationalists: in
the wake of their social ascendance they isolated as deviant and established as subjective those applications of phrenology which threatened
the industrial order and, thereby were able not only to promote as objectively grounded the class structures and values emerging as dominant, but also, at the same time, were able to subtly elevate in their
audiences' minds the idea that there actually was, at root, a pure

phrenology. In the name of respectability or 'purity', which Mary
Douglas well advises us to view in terms of the establishment of social
boundaries,[55] phrenologists in identifying a vulgar phrenology were
holding up to their audiences a species of scientific pornography.
They thus flattered their audiences' and their own integrity, objectivity and elitism and accomplished for phrenology's social and scientific
respectability what real pornography did for respectable Victorian society: that is, creating what Steven Marcus has called a 'negative
analogue'.[56] Given that phrenology was the most popular and popularized science in the early nineteenth century, it is not unwarranted to
claim that in emphatically railing against this 'negative analogue' in
order to overcompensate their pseudoscientific vulnerability, phrenologists, ironically, did more than any other single group to spread the
positivist image of the neutrality of science and of the legitimacy of
the dichotomies fact/value, science/ideology, science/pseudoscience.

Needless to say, the device of drawing their audiences' attention
to the 'polluters' of their science had the intended effect of helping
to obscure the claims of the antiphrenologists. But the main point I
want to establish here is the general one of how scientisms (so-called)
are created for the purpose of mystifying the actual social origins of
a science. Through this Dickensian caricature of a science we can
clearly see how the false reification science/scientism or science/
pseudoscience is actively created by persons involved with asserting
specific social relations and how that reification, once established,
functions in such a way as to mystify and sustain those social relations. For the Industrial Revolution, then, phrenology graphically re-
hearses the same process that (according to Mendelsohn) was conducted
by the second generation in the Scientific Revolution of the seventeenth century[57] immediately after the birth of capitalism as a political system.

(iv)

But this is not the end of the story. What I have just sketched
was historically specific to a stage of socio-economic and cultural
transition from pre- to post-reformed industrial bourgeois society.
Quite as revealing for our thesis is what occurred to phrenology after
this period when, as an elitist movement it went into decline. According to Whig historiography this phase of phrenology's history

represents the inevitable triumph of value-free science over that which had always been socially and politically polluted.[58] And here too, the factual basis for the Whig interpretation is compelling: one scarcely needs to look further than the local phrenological societies in which by the late 1830s medical membership had drastically fallen off, or to the medical journals where there was, if not complete agreement with the view that phrenology was a "monstrous abortion", at least a decided cooling of the earlier enthusiasm.

But the view that this therefore represents the triumph of scientific reason is no less of a loaded assumption than the view that people's initial interest in phrenology was governed by the irrational. As with the rise of phrenology so with its decline, a social interpretation does better justice to the facts. For one thing it recognizes that by the 1840s the period of transition was passed and that the new alignments of social power for which phrenology had served were becoming firmer. With phrenology's progressive role receding, it was no longer as necessary to exploit the vulgarizers of the science for that end. Indeed, phrenology itself might now come to serve the role intended by the 'negative analogue' of its own creation. If identified as 'pseudoscience' it might serve the new stabilized order as it had served the social order of the 1820s, that is as an agent for the intellectual production of the reified separations (fact/value, object/subject, body/mind, nature/culture, science/society) that preserve a socio-economic world of manipulation, domination and exploitation both of natural processes and social relations.[59] It must be said, however, that this view of phrenology's role is one that is gained only in hindsight, being culled more from the late-Victorian positivist historiography than from the actual history of the mid-Victorian period.[60] To liberal contemporaries it was enough to recognize that phrenology as a social resource was simply no longer useful in the ways it had been in the past. It should be said as well that, depending largely on socially-governed self-perception and on past degrees of commitment, different individuals came at slightly different times to make this recognition of phrenology's inutility. Thus as early as 1836 a reviewer in one of the Unitarian journals could reflect:

> Phrenology fairly ran away with us in our youth;
> but she carried us over ground so rugged and into
> mists so thick that as we grew stronger we grew
> desirous of emancipation, and at last burst away
> out of her labyrinthine recesses.[61]

For those who had been deeply involved in phrenology's propagation,
though, it was less easy to 'burst away' and still keep face. Far more
characteristic was the gradual withdrawal of acknowledged interest such
as can be seen in the *Lancet* in the late 1830s where, it is also significant, new forms of dissenting knowledge, homoeopathy and mesmerism in
particular, began to come under attack as socially-infiltrated 'pseudo-
sciences'.[62]

A specific date can be placed on this creeping awareness of phrenology's inutility and its social determinism can itself be confirmed
through the important article on "True and False Phrenology" which appeared in the *British and Foreign Medical Review* of July 1842. In this
it was made quite explicit that a watershed had been reached between a
popular radical phrenology and a socially-respectable scientific phrenology and that the time had come either to purge the science of its
"half-proved truth and empty conjecture" and its unbecoming popular
trappings that arrested its scientific progress, or to abandon it altogether. Assumed by many to have been written by the editor of the journal, John Forbes, who up to this time had been a leading member of the
Phrenological Association, the article was in fact written by a colleague and alter-ego of his, Dr. Daniel Noble. Noble too up to this
time had been a leading promoter of phrenology; he was the President of
the Manchester Phrenological Society between 1835 and 1838 and, whilst
strongly campaigning for medical reforms at local and national levels
in public and institutional forms, he had written dozens of papers supporting phrenology and had lectured publicly on the science in the
Manchester 'Lit & Phil'. By the 1840s Noble had established himself as
one of Manchester's leading physicians, consultants and medical lecturers, he had acquired a national reputation as an authority on psychological medicine, was President of the Lancashire and Cheshire
branch of the British Medical Association and was even offered a County
Magistracy. In the 1850s he was to be elected a Fellow of both the
Royal College of Physicians and the Royal College of Surgeons.

His "True and False Phrenology" thus coincides with his rise in
the social hierarchy, with his newly found status, power and wealth.
It also coincides with what had become abundantly clear at the Phrenological Association meeting of the month before, that phrenology had
little hope left of ever becoming an official section of the British
Association for the Advancement of Science. Noble and Forbes were
among those who dissented strongly from the radical materialist views

put forward at that meeting by the London phrenologists under Elliotson aegis,[63] for they understood that such views would make it impossible for the British Association to give phrenology the *social* sanctification that would come with the scientific recognition. As Noble confessed in his article, it was not the scientific problems of phrenology that made it reprehensible to the educated, but the fact that it was too obviously continuing (especially among the "facile" and "superficial" multitude) to legitimate radical anti-establishment ideas which seriously compromised the coherence of the bourgeois industrial order. With a strong sense of the stiffening social relations of industrial society and the security of his own elitist position, Noble no longer wished to be reminded that he had acted the leveller's part in once securing for the Manchester Phrenological Society a cast of the head of the radical William Cobbett. Noble may have believed, as many other phrenologists did, that the science could still be of essential service in manipulating working-class opinion so as to protect the new social hierarchy. But set against this was an awareness that the cohering ideological function of the science was blighted (along with his own professional prestige) if phrenology was rendered the tool of divisive political creeds. If phrenology could not be retained as ostensibly value-neutral, then Noble was prepared to abandon it on grounds of scientistic overkill. A few years later he followed his own counsel and opted for the respectable brain physiology of his major critic, William Carpenter.[64]

Another who further illustrates the socially and politically defensive reasons behind phrenology's decline as an elitist movement is the botanist Hewett Watson. Like Noble, he also had been among the vanguard of the phrenological counter-culturalists, but in 1840 he resigned as editor of the *Phrenological Journal* suddenly convinced "that much is stated in the writings of phrenologists which is doubtful, if not erroneous".[65] Watson gave no explanation of why it had taken him until 1840 to realize the science as 'pseudo', after having led the phrenological attack on Roget and Sir William Hamilton in the 1830s.[66] But his rebellion from his father's Toryism and his dissent from the Anglican Church in which he was raised tell us on the one hand as much about his involvement with phrenology, as his application for the Chair of Botany in Ireland in 1846 tell us about his reasons for withdrawing on the other.[67]

The same point can be repeated through the individual biographies of many other phrenologists. For a final and more collective illustration, however, let me briefly review the situation surrounding the establishment in December 1845 of a lectureship in phrenology at the Andersonian University. In setting this up (through funds provided by the will of one of the earlier wealthy patrons of phrenology) it was anticipated that practitioners then "between the ages of forty and fifty" who had sought the advantages of phrenology in the 1820s and 1830s would be anxious to send their sons through such a course.[68] But that the lectureship was to be almost wholly a tribute to the science's past relevance became clear when only twelve students enrolled and the course was forced to terminate after two sessions. Standing socially on much firmer ground than many of their fathers had, the sons (along with the fathers) could by 1845 nod sagely in agreement with the much-quoted statement of Benjamin Travers (Surgeon Extraordinary to Victoria and Albert) that phrenology owed its existence "to the countenance which it derives from a twilight of truth, though only sufficient to serve as a beacon to the absurdities with which it is enveloped".[69] In light of what I have been illustrating in this paper, it should be clear that only a determined Comtean Positivist could take the endorsement of such a statement among former supporters of phrenology as separate from those supporters' changed network of social relations or as separate from their cultural achievements. It makes far better sense to see the irrelevance of phrenology to its former advocates (and hence its freedom to be newly deployed as a pseudoscience in society generally) as a part of the same social process that had led to the rise of popular bourgeois phrenology. What had yet to be secured by the liberal bourgeoisie in the 1820s and what, with the most amazing social tensions and contradictions was sustained in the 1830s, was by and large accomplished by the 1840s.

(v)

To sum up: I've been arguing that the deployment of the label 'pseudoscience' always acts to conserve because, almost by definition it seems, the label appertains to perceived forms of social deviance or to what at any moment in history may be deemed morally illegitimate. To make my point I have ridden two horses into areas where conservatism in the deployment of pseudoscience is not readily apparent:

respectively, the socialist attack on IQ testing, and the early nineteenth century attack on social and scientific elitism by radical/lib-
eral phrenologists. I have argued that in both these cases the deployment of the label 'pseudoscience' acts conservatively because, unwittingly, it legitimates the bourgeois worldview that is mystified and
mediated through science. To identify and attack 'pseudoscience' is
both to reproduce the social and ideological structures and relations
of capitalist society and to reproduce the process of concealment originally conducted in the seventeenth century isolation as and exploitation of 'pseudoscience'. Through the deployment of 'pseudoscience'
the partial view of reality asserted by modern science remains thoroughly mystified under the metaphysic of 'truth' and 'certainty'.

Because all post-seventeenth-century attacks on 'pseudoscience'
must similarly conserve or protect the bourgeois ideology embedded in
science, it does not matter whether the deployment comes from the left
or from the right or whether the social movements which frequently toss
up the stipulation 'pseudoscience' have explicitly or implicitly to conserve a great deal or very little. My political horse rides as easily
through fields of reaction (eg. the antiphrenologists' deployment of
'pseudoscience') as through fields of sweeping social change. What one
sees overall through the socio-political vicissitudes of phrenology's
history in the nineteenth century is how new demands upon the bourgeois
order call forth new deployments of 'pseudoscience' which work either
overtly or covertly in the interest of conserving that order.

My other horse emerges from the stalls of the sociology of scientific knowledge, is jockied by 'relativism' (but see note 17) and is
for the most part an apolitical beast. Those disinclined to back the
political horse would do as well, if not better, to back this one which
is not dependent upon the neo-marxist critique of scientific objectivity. The reason for doing as well with this horse is simple: since all
knowledge of external nature is made by men and socially constructed,
the identification and criticism of any particular body of knowledge
as 'pseudoscientific' must count as a defence of some other body of
knowledge. In the case of phrenology, as in the contemporary situation
with IQ testing, claims and counter-claims of 'pseudoscientificity'
(i.e. claims and counter-claims of the social infiltration of the
other's knowledge) constitute the defence of human and social interests
in the name of an 'objective world' (however politically or otherwise
defined). To label 'pseudoscience' in the interest of creating this

'objective world' must be seen as deeply conservative, for it is a constraint on man's creative activity - a corset on any alternative view
of man making, not only his own history, but also his own reality.[70]

Thus, it seems to me, both horses pay to win in the argument that
the deployment of 'pseudoscience' always acts conservatively. If you
like, it is a photo-finish at the line or edge of objectivity.

Whatever may be felt about the nature of the argument here, I hope
that under the pretext of my simple thesis I have at least helped to
illustrate how bodies of scientific knowledge are criticised by identifying the presence of social interests in them and how that identification is essential to stipulating that certain bodies of knowledge are
not knowledge at all, but are 'ideology', 'pseudoscience', 'error',
etc.[71] It follows that the study of 'pseudoscience' has value not for
its own sake nor for the sake of legitimating as 'objective' what is
actually political in modern science, but for the sake of revealing
more about the science and society that negotiationally *defines* pseudo-
science through its interactions with it - both past and present. It
is only through this kind of understanding that we can hope to remove
the iniquities that are falsely and misleadingly attributed to 'pseudo-
science'. As a practical concluding remark, let me say that if the historical aim is description rather than prescription it would be infi-
nitely preferable to have the term 'pseudoscience' replaced in our vo-
cabularies with something like 'unorthodox science' or 'non-establish-
ment science', to imply degrees of difference in scientific social perception rather than categorical methodological opposites or matters of
truth and falsehood. If we must use the label 'pseudoscience' then let
us at least acknowledge that that usage, whether from the left or from
the right, is historiographically as ideological and covertly mystifying as its usage by any figure in history. For what does the focus on
'bad' science accomplish, but the obfuscation of the ideological power
of 'good' science? The underlying purpose of this paper has been to
demonstrate that the task before the historian of 'pseudoscience' is
not the reproduction of this essentially political process but its
avoidance and exposure.

Acknowledgements

I am grateful to Steven Shapin for his comments on an earlier version of this paper.

I would also like to specially thank the Calgary Institute for the Humanities for sponsoring this paper's preparation and the larger study upon which it is based.

FOOTNOTES

1   Everett Mendelsohn, "The Social Construction of Scientific Know-
    ledge", in Mendelsohn, Peter Weingart and Richard Whitley, eds. *The
    Social Production of Scientific Knowledge* (Dordrecht, Holland and
    Boston: D. Reidel, 1977), 3-26, esp. pp. 17-20.   *Cf.* Joseph Ben
    David's idealist account of the origins of seventeenth-century sci-
    ence in his *The Scientist's Role in Society* (Englewood Cliffs,
    N. J.:   Prentice-Hall, 1971) as criticized in Thorvald Gran, "Ele-
    ments from the debate on science in society: a study of Joseph
    Ben-David's theory", in Richard Whitley, ed. *Social Processes of
    Scientific Development* (London: Routledge, 1974), pp. 195-209, esp.
    pp. 206ff.   See also David Dickson, "Science and Political Hegemony
    in the 17th Century", *Radical Science Journal*, 8 (1979), 7-37.

2   That our modern scientific mentality stems from the seventeenth
    century is not of course a new idea; it was well understood by
    Lukács in the 1920s (see below note 15) and testified to by Herbert
    Butterfield in his Cambridge lectures of 1948 published as *The
    Origins of Modern Science* (London: G. Bell, 1949).

3   Steven Rose, "Scientific Racism and Ideology: the IQ racket from
    Galton to Jensen", in Hilary and Steven Rose, eds. *The Political
    Economy of Science: ideology of/in the natural sciences* (London:
    MacMillan, 1976), 112-41.

4   Barry Barnes, *Scientific Knowledge and Sociological Theory* (London:
    Routledge, 1974), pp. 132-36.

5   Les Levidow, "A Marxist Critique of the IQ Debate", *Radical Science
    Journal*, 6/7 (1978), at pp. 22-56.

6   Jeffrey Blum, *Pseudoscience and Mental Ability: the origins and
    fallacies of the IQ controversy* (New York: Monthly Review Press,
    1978), pp. 12-13.

7   *Ibid.*, p. 19.

8   Barnes, *op. cit.*, p. 134.

9   See Susan Cannon, "Professionalization", in her *Science in Culture:
    the early Victorian period* (New York: Dawson, 1978), 137-65.

        To anticipate the critique that follows, note that Blum's de-
    fence of 'real science' portrayed as collective enterprise, is a
    recapitulation of Bacon's ideal of scientific organization and is
    therefore, unbeknown to Blum, a defence of capitalist practices and
    forms over alternative practices and forms.   On the formal similar-
    ity between Bacon's scientific method and the capitalist labour
    process see Dickson *op. cit.* (note 1), pp. 11-12.

10 Blum, Ch. 12, "The Ideological Dimension", pp. 161-81.

11 Such a usage, derived from Althusser, may be found in Mary Hesse,
    "Criteria of Truth in Science and Theology", *Religious Studies*, 11
    (1975), 396. My own usage is derived more from Henri Lefebvre,
    "Ideology and the Sociology of Knowledge", in his *The Sociology of
    Marx*, trans. Norbert Guterman (Harmondsworth: Penguin, 1966), pp.
    59-88 and repr. in Janet L. Dolgin, D. S. Kemnitzer and D. M.
    Schneider, eds. *Symbolic Anthropology* (New York: Columbia Univer-
    sity Press, 1977), pp. 254-69; Clifford Geertz, "Ideology as a Cul-
    tural System", in David E. Apter, ed. *Ideology and Discontent*
    (Glencoe: Free Press, 1964), pp. 47-76; and the writings of Bob
    Young, of which see: "Science *is* Social Relations", *Radical Sci-
    ence Journal*, (1977), pp. 65-129, which contains a useful biblio-
    graphy of sources on science and ideology, including Young's several
    other works.

12 Throughout this paper 'scientism' refers to the transformation of
    positivism (the natural philosophy that combines the traditions of
    empiricist and rationalist thought) into social philosophy on the
    basis of which society is explained and interpreted. Scientisms
    are extrapolations from well-established scientific domains to so-
    cial domains, but since science itself is constructed under social
    conditions and subsumes these conditions, scientisms may be said
    only to accomplish overtly for social legitimation what sciences do
    covertly. See: Luke Hodgkin, "A Note on 'Scientism'", *Radical
    Science Journal*, 5 (1977), 8; David Dickson, "Technology and Social
    Reality", *Dialectical Anthropology*, 1 (1975), 34-37. *Cf*. the ortho-
    dox distinction between science and scientism in F. V. Hayek,
    "Scientism and the Study of Society", in his *The Counter-Revolution
    of Science. Studies on the abuse of reason*, (Glencoe: Free Press,
    1955), pp. 129-42; George Eastman, "Scientism in Science Educa-
    tion", *The Science Teacher*, 36 (1969), 19-22; and Robert B. Fisher,
    "Science and/or Scientism", *Science, Man and Society* (Philadelphia;
    W. B. Saunders, 1971), pp. 43-44.

13 Blum, Ch. 3, "Social Darwinism and the Eugenics Movement", pp. 34-
    42; and on Lysenko, pp. 147-60 *et passim*. Blum appears oblivious
    to the well-known and much-debated tautological problem involved in
    the use of the term 'Social Darwinism' which is contingent on the
    fact that Darwin derived the key to his transmutation thesis from
    the political economics of Malthus. See Robert M. Young, "Malthus
    and the Evolutionists: the common context of biological and social
    theory", *Past & Present*, 43 (1969), 109-45; Raymond Williams,
    "Social Darwinism", in Jonathan Benthall, ed. *The Limits of Human
    Nature* (London: Allen Lane, 1973), pp. 115-30; and, for a resumé
    and analysis of the historiographic debate, Steven Shapin and Barry
    Barnes, "Darwin and Social Darwinism: purity and history", in
    Barnes and Shapin, eds. *Natural Order: historical studies of Sci-
    entific Culture* (London/Beverly Hills: Sage, 1979), pp. 125-42.

    For the re-interpretation of Lysenkoism outside the use/abuse
    framework relied upon by Blum, see Dominique Lecourt, *Proletarian
    Science? The case of Lysenko*, with an introduction by Louis Al-
    thusser, trans. Ben Brewster (London: New Left Books, 1977); and
    Bob Young, "Getting Started on Lysenkoism", *Radical Science Journal*,
    6/7 (1978), 81-105. See also Philip Boys, "Detente, Genetics, and
    Social Theory", *ibid*, 8 (1979), 61-90; and Arthur Koestler, *The
    Case of the Midwife Toad* (London: Hutchinson, 1971).

14 Blum, pp. 163 ff.

15 George Lukács, *History and Class Consciousness* (1923; new ed. London: Merlin, 1971), pp. 7-11, 89-98 *et passim*. The relevant passages are cited in Gareth Stedman Jones, "The Marxism of the Early Lukács", *New Left Review*, 70 (1971), repr. in Stedman Jones, ed. *Western Marxism, A critical Reader* (London: New Left Books, 1977), pp. 13-14. See also Robert M. Young, "The Historiographic and Ideological Contexts of the Nineteenth-Century Debate on Man's Place in Nature", in Young and Mikuláš Teich, eds. *Changing Perspectives in the History of Science* (London: Heinemann, 1973), pp. 398-99, 402, 405, 414, 430-34. (Note: While I have used 'objectification' as a shorthand for reification, it should properly be understood as implying the process whereby the history of a thing is separated from its human origin and quality, i.e., as 'misplaced concreteness' or the assuming of entity existence of that which is actually 'metaphysical'.)

16 *Cf.* Mary Douglas, "Environments at Risk" in her *Implicit Meanings: essays in anthropology* (London: Routledge, 1975), pp. 230-48 and David Ingleby, "Ideology and the Human Sciences: some comments on the role of reification in psychology and psychiatry", in Trevor Pateman, ed. *Counter Course: a handbook for course criticism* (Harmondsworth: Penguin, 1972), pp. 51-81 (esp. at p. 54). Both amplify the point made by Roland Barthes: "To instil into the Established Order the complacent portrayal of its drawbacks has nowadays become a paradoxical but incontrovertible means of exalting it." "Operation Margarine" in his *Mythologies*, trans. Annette Lavers (St. Albans, Herts: Paladin, 1976), p. 41

    Lest there be any question, I am *not* saying that science is nothing but ideology, only that science is ideology *as well as* whatever else it is. If it is universally true that $E=mc^2$, it will remain true whatever the dominant ideology in any society.

17 See: David Bloor, "Wittgenstein and Mannheim on the Sociology of Mathematics", *Studies in the History and Philosophy of Science*, 4 (1973), 173-91; H. M. Collins and Graham Cox, "Recovering Relativity: did prophecy fail?", *Social Studies of Science*, 6 (1976), 423-44; H. M. Collins, "The TEA Set: tacit knowledge and scientific networks", *Science Studies*, 4 (1974), 165-86; *idem*, "The Seven Sexes: a study in the sociology of phenomenon, or the replication of experiments in physics", *Sociology*, 9 (1975), 205-24. Particularly relevant applications of the 'relativist' approach to the study of 'pseudoscience' (though unfortunately unavailable to me during this paper's preparation) are H. M. Collins and T. J. Pinch, "The Construction of the Paranormal", and R. G. Dolby, "Reflections on Deviant Science", both in Roy Wallis, ed. *On the Margins of Science: the social construction of rejected knowledge*, Sociological Review Monographs (University of Keele: 1979).

    I have been mostly informed in this paper by the historical applications of ethnomethodology in the studies by Barry Barnes and Steven Shapin. They have relied heavily on the insights of anthropology (especially that of Mary Douglas) to underline the fact that the question of truth and falsehood in science is neither here nor there: 'truth', they argue, is relative to the social context of its use, and to question it historically is to avoid the more important problem of how science functions in society. "Truth", as Mary

Douglas reminds us in the words of Sir James Frazer, "is only the
hypothesis which is found to work best." *Purity and Danger: an
analysis of concepts of pollution and taboo* (London: Routledge,
1966), p. 24 quoted in Shapin, "Homo Phrenologicus: anthropological perspectives on an historical problem", in Barnes and Shapin,
op. cit. (note 13), p. 44. (For the works of Barnes and Shapin see
the bibliographies in *Natural Order*.)

> For general criticism on the relativist position (on the
grounds that not *all* truths are context-dependent because some [à
la Pareto's 'residues'] are universal and fundamental) see Steven
Lukes, "On the Social Determination of Truth" and "Relativism:
cognitive and moral", both in his *Essays in Social Theory* (London:
Macmillan, 1977); *idem*, "Some Problems About Rationality", in
Bryan R. Wilson, ed. *Rationality* (Oxford: Blackwell, 1970), pp.
194-213; and Jack D. Douglas, "Understanding Everyday Life" in
Jack D. Douglas, ed., *Understanding Everyday Life* (Chicago: Aldine,
1970), pp. 3-44, esp. at pp. 39ff.

> My own position is that what is *argued* as scientific 'truth'
or 'certainty' is context-dependent, (is invented, not discovered),
but there is nevertheless a real world of non-ideological facts
(external nature) outside one's consciousness (as Feuerbach argued
against Hegelian idealism). I agree with Marx, that "The *human*
significance of nature only exists for *social* man" and therefore
only through social change is the human significance of nature
changed. (For relevant sources on this philosophical problem see
Melvin Rader, *Marx's Interpretation of History* [New York: Oxford
University Press, 1979], pp. 108-9 and Alfred Sohn-Rethel, "Science
as Alienated Consciousness", *Radical Science Journal*, 2/3 [1975],
p. 80n. See also Raymond Firth, "The Sceptical Anthropologist?" in
Maurice Bloch, ed. *Marxist Analysis and Social Anthropology* [Lon-
don: Malaby Press, 1975], pp. 29-60 [esp. at p. 31]; and Karl
Figlio, "Chlorosis and Chronic Disease in Nineteenth-Century Britain: the social constitution of somatic illness in a capitalist
society", *Social History*, 3 [1978], 168).

18 For a review of the recent literature on phrenology and a fairly
complete bibliography of past writings on the subject see my
"Phrenology: the provocation of progress", *History of Science*, 14
(1976), 211-34. For the basic premises of the science see the
paper by Trevor Levere in this volume.

19 Karl M. Dallenbach, "Phrenology versus Psychoanalysis", *American
Journal of Psychology*, 68 (1955), 511-25; P. B. Medawar, "Further
Comments on Psychoanalysis" in his *The Hope of Progress* (London:
Methuen, 1972), pp. 57-68.

20 Rose, *op. cit.* (note 3), pp. 117, 129.

21 Hans Jurgen Eysenck, *Fact and Fiction in Psychology* (Harmondsworth:
Penguin, 1965); pp. 130-131 (where the essay by Dallenbach cited
above comes in for praise); *Sense and Nonsense in Psychology* (Harmondsworth: Penguin, 1958), p. 61 (which in the same breath links
in mesmerism); and *Uses and Abuses of Psychology* (Harmondsworth:
Penguin, 1954), pp. 28-29.

22 See, in particular, David de Giustino, *Conquest of Mind, Phrenology and Victorian social thought* (London: Croom Helm, 1975); John D. Davies, *Phrenology, Fad and Science: a 19th-century American crusade* (New Haven: Yale University Press, 1955); Alastair Cameron Grant, "George Combe and His Circle: with particular reference to his relations with the United States of America" (Ph.D., Edinburgh, 1960); T. M. Parssinen, "Popular Science and Society: the phrenology movement in early-Victorian Britain", *Journal of Social History*, 7 (1974), 1-20; Angus McLaren, "Phrenology: medium and message", *Journal of Modern History*, 46 (1974), 86-97; Owsei Temkin, "Gall and the Phrenological Movement", *Bulletin of the History of Medicine*, 21 (1947), 102-15; (anon. for the present), "Phrenology in France", *Journal of the History of the Behavioral Sciences* (forthcoming); and, from a different and deeper social perspective, the papers by Steven Shapin cited below (note 32).

23 Robert Darnton, *Mesmerism and the End of the Enlightenment in France* (New York; Schocken, 1970).

24 None of this is any longer in doubt and is strongly reinforced by the evidence given in my "The Cultural Meaning of Popular Science: phrenology and the organization of consent in nineteenth century Britain" (Ph.D., Cambridge, 1978). Because this study is forthcoming by Cambridge University Press, and because the point of this paper is not to marshall evidence but to argue on the basis of that evidence, I have in what follows avoided where possible detailed citations. The evidence upon which the rest of this paper is based is taken from the first two chapters of my study.

25 The dismissive phase of Bentham's. *The Works of Jeremy Bentham*, John Bowring, ed. (1843), vol. 7, pp. 433-34.

26 The literature for the sociological sense of 'institutionalization' is cited in Steven Shapin, "Social Uses of Science, 1660-1800", in G. S. Rousseau and Roy Porter, eds. *The Ferment of Knowledge* (Cambridge: Cambridge University Press, forthcoming).

27 [Hewett Cotrell Watson], *Phrenological Journal*, 11 (1838), 263.

28 See my "Phrenology and British Alienists, c. 1825-1845", *Medical History*, 20 (1976), 1-21, 135-51.

29 The specifics of the medical interest are given in *ibid.*, esp. pp. 13-21.

30 As put forward by Charles C. Gillispie, "The *Encyclopedie* and the Jacobin Philosophy of Science: a study in ideas and consequences", in Marshall Clagett, ed. *Critical Problems in the History of Science* (Madison: University of Wisconsin Press, 1959), pp. 255-89.

31 The literature delineating this 'collision' in Britain, or, more correctly, the transformation in the social use of science in Britain in the early nineteenth century, has rapidly expanded over the last decade. For useful introductions see Morris Berman, *Social Change and Scientific Organization: the Royal Institution, 1799-1844* (Ithaca, New York: Cornell University Press, 1978); *idem*, "'Hegemony' and the Amateur Tradition in British Science", *Journal of Social History*, 8 (1975), 30-50; Ian Inkster, "Science and

Society in the Metropolis: a preliminary examination of the social
and institutional context of the Askesian Society of London, 1796-
1807", *Annals of Science*, 34 (1977), 1-32; Steven Shapin, "The Pot-
tery Philosophical Society, 1819-35: an examination of the cultural uses of provincial science", *Science Studies*, 2 (1972), 311-36;
Arnold Thackray, "Natural Knowledge in Cultural Context: the Manchester model", *American Historical Review*, 79 (1974), 672-709.
See also the essays in J. B. Morrell and Ian Inkster, eds. *Metropolis and Province: British science 1780-1850* (London: Hutchinson, forthcoming); and my forthcoming "Refining the Social Land-
scape for Popular Science in Early Industrial Britain".

32 Such a map, based on 28 major medical supporters of phrenology in
Britain and 14 major scientific opponents, is discussed in Cooter,
"Cultural Meaning of Popular Science", *op. cit.* (note 24), Ch. 1,
pt. iv, with biographical details in Appendix A.

That the phrenology debates were about social and ideological
interests was first made clear in Steven Shapin, "Phrenological
Knowledge and the Social Structure of Early Nineteenth-Century
Edinburgh", *Annals of Science*, 32 (1975), 219-43. Since then
Shapin has moved beyond the prosopographical and rhetorical terrain
of the phrenology debates and into their scientific substance to
further illustrate the social and ideological strategies involved;
see Shapin, "The Politics of Observation: cerebral anatomy and social interests in Edinburgh phrenology disputes", in Wallis, *op.
cit.* (note 17), pp. 139-78. In "Homo Phrenologicus", *op. cit.*
(note 17), with a heavy reliance on anthropological writings,
Shapin has sought to work out "a social epistemology appropriate
for the history of science" by ascertaining through the historical
case of phrenology in the context of its use in Edinburgh the nature of the links between natural knowledge and social context;
while in his "'Merchants in Philosophy': the politics of an Edinburgh plan for the diffusion of science, 1832-1836" in *Metropolis
and Province*, *op. cit.* (note 31), he endeavours to illuminate the
political and logistic problems involved in the active diffusion of
scientific ideas by a careful local study of the involvements of
Edinburgh phrenologists in the Edinburgh Philosophical Association.

33 Karl Mannheim, "Conservative Thought" and "The History of the Concept of State as an Organism: a sociological analysis", both in
his *Essays on Sociology and Social Psychology*, ed. Paul Kecskemeti
(London: Routledge, 1953). For an illuminating deployment of the
concept of 'styles of thought' to a similar problem see Jonathan
Harwood, "The Race-Intelligence Controversy: a sociological approach", *Social Studies of Science*, 6 (1976), 369-94 and 7 (1977),
1-20.

34 It is difficult to confine Roget (1779-1869), for example, to an
aristocratic orbit. Though his uncle was Sir Samuel Romilly, and
though he became a Fellow of the Royal Society in 1815 and was the
Secretary and editor of the Society's proceedings between 1827 and
1848, he was by association and marriage more closely allied to
bourgeois mercantile sources of wealth and utilitarian philosophy.
He is more representative of the members of the Manchester Literary
and Philosophical Society or the Royal Institution, or the Society
for the Diffusion of Useful Knowledge, of which he was, respectively, a member, a Governor, and a committee member. He was intimate

with Bentham and other Philosophical Radicals; was Physician to the
Manchester Infirmary in 1805 and later to the Northern Dispensary
in London; and helped found the Manchester Medical School and the
University at London. See D. L. Emblem, *Peter Mark Roget: the
word and the man* (New York: Crowell, 1970); on the interests of
the members of the Manchester 'Lit & Phil' see Thackray, *op. cit.*
(note 31), and on the Royal Institution and the Society for the
Diffusion of Useful Knowledge see Berman, *op. cit.* (note 31), p. 112
et passim.

35   See Robert Mcaulay, "Velikovsky and the Infrastructure of Science:
     the metaphysics of a close encounter", *Theory and Society*, 6 (1978),
     313-42.   Useful parallels can also be drawn from the literature on
     the early nineteenth-century debate in geology between catastro-
     phists and uniformitarians; from that on the early twentieth-cen-
     tury disputes between the 'biometricians' and 'Mendelians' and be-
     tween the Darwinians and Lamarckians; and from the literature on
     the current debate between hereditarians and environmentalists over
     IQ. See, respectively, Walter F. Cannon, "The Uniformitarian -
     Catastrophist Debate", *Isis*, 51 (1960), 38-55; Donald MacKenzie and
     Barry Barnes, "Scientific Judgement: the biometry-mendelism con-
     troversy", in Barnes and Shapin, *op. cit.* (note 13), pp. 191-210,
     Koestler, *op. cit.* (note 13); and Harwood, *op. cit.* (note 33), and
     idem, "Heredity, Environment, and the Legitimation of Social Poli-
     cy", in Barnes and Shapin, *op. cit.* (note 13), pp. 231-51.

36   *Cf.* G. N. Cantor, "The Edinburgh Phrenology Debate: 1803-1828,"
     Annals of Science, 32 (1975), 195-218 and *idem.*, "A Critique of
     Shapin's Social Interpretation of the Edinburgh Phrenology Debate",
     ibid., 32 (1975), 245-56.

37   "Review of *Illustrations of Phrenology*....by Sir G. S. Mackenzie,
     Bart.", *Edinburgh Monthly Review*, 5 (1821), 94, cited in Shapin,
     "Phrenological Knowledge", *op. cit.* (note 32), p. 234. (The author
     of the review was likely the editor, Richard Poole, himself a mem-
     ber of the Phrenological Society of Edinburgh and the editor of the
     first four numbers of the *Phrenological Journal*.)

38   See "Cruelty to Animals - Sir William Hamilton's Experiments", *Phre-
     nological Journal*, 7 (1831-32), 427-33. The political/philosophi-
     cal/social meaning of vivisection in the early nineteenth century
     is discussed in Owsei Temkin, "Basic Science, Medicine, and the
     Romantic Era", *Bulletin for the History of Medicine*, 37 (1963), 97-
     129, repr. in Temkin, *The Double Face of Janus* (Baltimore/London:
     Johns Hopkins University Press, 1977), pp. 345-372, esp. at pp. 364-
     65. See also Richard French, *Antivivisection and Medical Science
     in Victorian Society* (Princeton: Princeton University Press, 1975),
     esp. Ch. 2.

     Phrenology's methodological conformity with the image of sci-
     ence meant that little reference could be made to it by contempor-
     aries writing on the difference between 'genuine science' and
     'pseudoscience'. See, for example, G. Cornewall Lewis, *An Essay on
     the Influence of Authority in Matters of Opinion* (London, 1875
     [1849]), p. 53; [Elizabeth Eastlake], "Physiognomy", *Quarterly Re-
     view*, 90 (1851), 62; and Henry Dircks, *Scientific Studies: II
     Chimeras of Science: astrology, alchemy, squaring the circle, per-
     petuum mobile, etc.* (London, 1869), esp. pp. 41, 80.

39 For sources see note 32, especially, "Phrenological Knowledge", pp. 231-43.

40 The rule itself could be discredited by quoting Bacon, as in the following from a Birmingham phrenologist attacking the Oxbridge academic elite: "To use the words of Lord Bacon, 'the exercise' which these Oxonians fulfil, 'fitteth not the practice or the image of life,' but 'doth pervert the motions and faculties of the mind, and not prepare them.'" He also accused the Oxonians of being priest-ridden. [Joshua Toulmin Smith], "On the Progressive Diffusion of Phrenology", *Phrenological Journal*, 10 (1836-37), 405. Bacon, as Gillispie has pointed out and as Roszak has amplified, was not a prophet of science so much as a prophet of an *image* of secular scientific and social progress, which has always been far more popular than science has been. Theodore Roszak, *Where the Wasteland Ends* (New York: Doubleday, 1972), pp. 148-49. Dickson *op. cit.* (note 1), has gone much further in asserting that the attraction of Bacon's philosophy "lay not merely in the fact that it offered a method for achieving useful knowledge, but that this method was a blueprint of the capitalist labour process, both in the way that [Bacon] suggested science as a social activity should be carried out, and in the particular forms in which [his philosophy] presented both things and people." (p. 11). *Cf.* on 'Baconism', Cannon, *op. cit.* (note 9), pp. 59, 73, 228-29ff.

41 As Alan Macfarlane has recently reminded us, the origins of capitalism extend much further back than the seventeenth century: *The Origins of English Individualism* (Cambridge: Cambridge University Press, 1978). However, as Dickson, *op. cit.*, points out, "What was significant about the political changes of the seventeenth century...was that they expressed the emergence into a position of dominance of a class defined by its role as controllers of the labour process." (p. 10).

42 This is not of course the only way in which we may see the phrenologists as apologists for a bourgeois urban industrial society. One must also look at their knowledge in the later context of its use, perceiving, for example, how their empiricist stress on perception led to an abdication of reflection and an absolutizing of positivist 'facts' that would reify the reformed status quo among popular audiences. More superficially, as Cannon says of all bearers of eighteenth-century utilitarianism, phrenologists were simply good conservative reformers. Cannon, *op. cit.* (note 9), p. 59.

43 On the functionary role of intellectuals generally, see Antonio Gramsci, "The Intellectuals", in *Selections from the Prison Notebooks*, ed. and trans. by Q. Hoare and G. Norwell Smith (London: Lawrence and Wishart, 1971), pp. 3-23.

44 All of which evidence will be seen as also confirming the view of Marx and Engels: "What else does the history of ideas prove, than that intellectual production changes its character in proportion as material production is changed?" *Manifesto of the Communist Party* (1848) in *Collected Works*, 6 (London: Lawrence and Wishart, 1976), 503.

45 *Examiner*, 30 October 1831.

46 "[Review of] Combe's *Constitution of Man*, 4th edition", *British and Foreign Review, or European Quarterly Journal*, 12 (1841), 142.

47 This should not be interpreted as meaning that the public necessarily deferred to medical opinion. Hostility, particularly among the working class, was widespread as is testified to by the popularity of unorthodox medical knowledge and therapeutics - eg. homoeopathy, hydropathy and patent medicines.

48 *Gentleman's Magazine*, 89 (Supplement, 1819), 608-9; see also *ibid.*, 93 (1823), 149-51 and 94 (1824), 301-3 and *cf*. with their earlier attitude, eg., 76 (1806), 502.

49 Daniel Ellis, *Memoir of the Life and Writings of John Gordon M.D., F.R.S.E. Late Lecturer on Anatomy and Physiology in Edinburgh* (Edinburgh, 1823), p. 42.

50 See, for eg., the surviving outline of Spurzheim's lecture course at the London Institution in 1827, reproduced in Cooter, "Cultural Meaning of Popular Science", *op. cit*. (note 24), following p. 59.

51 *Cf*. Barthes, "The Brain of Einstein", in his *Mythologies, op. cit.* (note 16), pp. 68-70, from whom the analogue with $E=mc^2$ is drawn.

52 *Cf*. Christopher Hill, "William Harvey and the Idea of Monarchy", *Past & Present*, 27 (1964), 54-72 repr. in Charles Webster, ed. *The Intellectual Revolution of the Seventeenth Century* (London: Routledge, 1974), pp. 160-81; and Roy Willis, *Man and Beast* (London: Hart-Davis MacGibbon, 1974). See also Steven Shapin and Barry Barnes, "Head and Hand: rhetorical resources in British pedagogical writing, 1770-1850", *Oxford Review of Education*, 2 (1976), 231-54.

53 This is well illustrated in Shapin, "Homo Phrenologicus" *op. cit.* (note 17).

54 While this need became generally greater for all respectable phrenologists in the 1830s, it could be as strongly felt by some phrenologists much earlier, depending on the nature of their social orbits. The phrenologist Sir George Mackenzie, for instance, who moved in aristocratic circles (though he was largely alienated from the social consciousness of his class), can be quoted as early as 1820 saying "it is *impossible* to know, by external signs alone, the *character* of any individual": *Illustrations of Phrenology* (Edinburgh, 1820), p. 25, cited by Cantor *op. cit.* (note 36), p. 252 in a mistaken attempt to discredit Shapin on the empiricism of phrenologists.

A similar contradiction appears to have forced some of the writers for the Society for the Diffusion of Useful Knowledge: on the one hand they wanted to keep their science lessons simple in order to convey certain values to the working class, but on the other hand they did not want to lose respect among scientific peer groups who might label them declamatory and unscientific. Such a problem obviously confronted the botanist John Lindley: see J. N. Hays, "Science and Brougham's Society", *Annals of Science*, 20 (1964) 239.

55 Mary Douglas, *Purity and Danger*, *op. cit.* (note 17). See also Mendelsohn on this in his "Social Construction of Scientific Knowledge", *op. cit.* (note 1), p. 7 and n. 17 p. 22. See also the review of Douglas' work (specifically *Implicit Meanings*) in relation to its value for the history of science by Barry Barnes and Steven Shapin, "Where is the Edge of Objectivity?", *British Journal for the History of Science*, 7 (1977), 61-66.

56 Steven Marcus, *The Other Victorians* (London: Weidenfeld and Nicholson, 1966), p. 283. An example of the effect is provided in the *Elgin Courier*, which commented upon the vulgarizing itinerant phrenologist J. Boyd, "It is grievous to find such attempts to destroy and render ridiculous an important science." 24 July 1846, quoted in *Phrenological Journal*, 19 (1846), 383. For references to other examples see Cooter, "Cultural Meaning Popular Science", *op. cit.* (note 24), p. 345 ff.

57 Mendelsohn, "Social Construction of Scientific Knowledge", *op. cit.* (note 1), esp. p. 17 ff.

58 Revisionist historians of phrenology do not quarrel with this interpretation. That further scientific research undermined phrenology's credibility is generally taken (contrary to the facts) as one of the reasons for its decline, along with the role of phrenological quackery, and an idea that phrenology ultimately spread itself too thinly (though not unimportantly) in too many directions at once. See De Guistino, *Conquest of Mind*, *op. cit.* (note 22).

59 See W. Leiss, *The Domination of Nature* (New York: Braziller, 1972), and Robert M. Young, "Why are Figures so Significant? The Role and the Critique of Quantification", in John Irvine, *et al* eds. *Demystifying Social Statistics* (London: Pluto Press, 1979), pp. 63-74.

60 Nor is the distinction between the early and later deployments of phrenology as straightforward as this would suggest. Especially over the latitude to be given to orthodox religion, reformers differed sharply. Hence, for example, the reform-minded and utilitarian but religiously orientated *Educational Magazine and Journal of Christian Philanthropy and of Public Utility* could claim of phrenology in May 1835 (p. 307) that it "is to the science of mind, what astrology was to astronomy or alchemy to chemistry; with this difference, that it seeks to mystify by assumed facts, instead of occult mystery."

61 *The Christian Teacher*, 2 (1836), 603-4.

62 See for example, W. C. Engledue, "Messrs. Forbes, Wakely, and Co, the Antimermeric Crusaders", *Zoist*, 4 (1847), 584-601; and Dr. Robert M. Glover, "Lectures on the Philosophy of Medicine, delivered in the Lecture Room of the Literary and Philosophical Society, Newcastle, Lecture VI: Quackery...Pseudo Science: Phrenology, Mesmerism, Hydropathy, Teetotalism, Vegetarianism, Homoeopathy", *Lancet*, 11 Jan. 1841, 34-38.

63 See W. C. Engledue, *Cerebral Physiology and Materialism, with the result of the application of animal magnetism to the cerebral organs. An address delivered to the Phrenological Association in*

London, June 20, 1842, with a letter from Dr. Elliotson, on mesmeric phrenology and materialism (London, 1842). Engledue was co-editor with Elliotson of the *Zoist*.

64   True confessions are given in Noble, *Elements of Psychological Medicine* (London, 1853), pp. x-xi, 36-38.

65   *Phrenological Journal*, 13 (1840), 387.

66   [H. C. Watson], *Strictures on AntiPhrenology, in two letters to Macvey Napier, Esq. and P. M. Roget, M. D. Being an exposure of the article called 'Phrenology', recently published in the Encyclopedia Britannica* ([for private distribution] Oct. 1838); *idem.*, "Phrenology and the *Encyclopedia Britannica*; or the deliberate obstruction of Truth", *Phrenological Journal*, 11 (1838), 278-82; and *idem.*, " Review of Sir William Hamilton's prefix to Alexander Munro's *The Anatomy of the Brain* (1831)", *ibid.*, 7 (1831-32), 427-33.

67   For biographical details on Watson see: "The Naturalist's Literary Portrait-Gallery No. III - Hewett Cottrell Watson, Esq. F. L. S.", *Naturalist*, 4 (1839), 264-69, and Frank N. Egerton III, "Hewett Cottrell Watson", *Dictionary of Scientific Biography*, 14 (New York, 1976), 189-91.

68   "Lectureship of Phrenology in the Andersonian University", *Popular Record of Modern Science*, 27 Dec. 1845, quoted in *Journal of Health and Disease*, 1 (1846), 232. See also, Andrew Combe, *Phrenology - Its Nature and Uses: an address to the students of Anderson's University at the opening of Dr. Weir's first course of lectures on phrenology in that institution January 7th 1846* (Edinburgh, 1846); *Lancet*, 10 Jan. 1846, 61-64; *British and Foreign Medical Review*, 22 (1846), 230-31; and Charles Gibbon, *The Life of George Combe* (London, 1878), vol. 2, 210-11.

69   Benjamin Travers, *The Physiology of Inflammation* (London, 1841), p. 44 and quoted in Sir George Lefevre, "Phrenology" in his *An Apology for the Nerves* (London, 1844), p. 58 and the *Medico-Chirurgical Review*, new series, 1 (1845), 369.

70   See Paul Feyerabend, *Against Method* (London: New Left Books,) 1975) and *Science in a Free Society* (London: New Left Books, 1978).

71   Shapin, "The Politics of Observation", *op. cit.* (note 32), p. 140.

APOCRYPHAL KNOWLEDGE: THE MISUSE OF SCIENCE

Consider the following story. There is a species of fish that in-
habits the Great Barrier Reef off the coast of Australia. These fish
travel in schools composed entirely of females with a dominant male
leader. Should some misfortune befall the male, one of the female fish
will take over the leadership of the school and become physiologically
transformed into a male. All will then proceed as before.[1] What conclusion can be drawn from the strange tale of these fish? In the modern world inhabited by Naked Apes who express Territorial Imperative
and are motivated by Selfish Genes, one would not be surprised to find
the following arguments based on the life-cycle of these Australian
fish. A feminist might sieze upon them as biological proof that women
are every bit as good as men and that, given appropriate environmental
changes, they can perform the same social tasks as men. On this basis
she would argue for equal rights for women in contemporary society.
A male chauvinist, on the other hand, might argue on the basis of the
same school of fish that they just go to show that if you let a woman
rise in the administration she will do so at the peril of losing her
femininity. Consequently, he would argue, women should have different
social roles from men.

Clearly, something is wrong with these arguments. In sound reasoning it is not possible to draw contradictory conclusions from ostensibly identical sets of premises. In this paper I will demonstrate
that there are several fundamental problems with such appeals to scientific facts. Although the story of the fish may appear trivial and
may indeed be apocryphal, it is paradigmatic of a style of reasoning
which has been common for over a century.

Examples of such attempts to use science to justify social or ethical programs abound. Perhaps the most dramatic example, because it so
clearly illustrates the viciously self-contradictory nature of such
reasoning, is the variety of uses to which Darwin's theory of evolution
has been put. Ever since Darwin used science to define human nature
biologically in *The Origin of Species* (1859) and *The Descent of Man*
(1871), social prescriptions based on Darwinian principles have pro-
liferated. Herbert Spencer (1820-1903), for example, developed a set
of sociological principles and prescriptions based on the evolutionary

concept of the survival of the fittest.[2] Arguing that any interference
with the natural course of events would lead to the survival of the un-
fit, Spencer opposed any interference with the "natural" course of social growth. Population pressure alone, he thought, would lead to social improvement. Thus he believed that any kind of social welfare
system would impede this natural, ameliorative process. Consequently
he opposed all state aid to the poor, state-supported education, sani-
tary supervision, housing regulation, licensing of physicians, and the
government postal service.[3] William Graham Sumner, an American disciple of Spencer's, eloquently summed up this view:

> The millionaires are a product of natural selection,
> acting on the whole body of men to pick out those
> who can meet the requirements of the work to be
> done....It is because they are thus selected that
> wealth - both their own and that entrusted to them -
> aggregates in their hands. They may fairly be re-
> garded as the naturally selected agents of society
> for certain work. They get high wages and live in
> luxury, but the bargain is a good one for society.
> There is the intensest competition for their place
> and occupation. This assures us that all who are
> competent for this function will be employed in it,
> so that the cost of it will be reduced to the low-
> est terms.[4]

Lester Ward (1841-1913), another founding father of American Sociology, found a very different use for biological science as applied
to the study of society. Regarding human mental capacities as the product of evolution, Ward urged the use of scientific knowledge of nature and society for the improvement of the human condition. "The day
has come," he wrote in *The Psychic Factors in Civilization* (1893), "for
society to take its affairs into its own hands and shape its own desti-
nies."[5]

> I regard society and the social forces as constitut-
> ing just as much a legitimate field for the exercize
> of human ingenuity as do the various material sub-
> stances and physical forces. The latter have been
> investigated and subjugated. The former are still
> pursuing their wild, unbridled course.[6]

Ward mocked Spencer and Sumner, who drew *laissez-faire* conclusions from
Darwinism, citing their inconsistency in accepting the technological
benefits of manipulating physical nature, while denying that any good
could come from analogous use of the scientific laws that govern the
social order.

> When a well-clothed philosopher on a bitter winter's
> night sits in a warm room well lighted for his pur-
> pose and writes on a paper with pen and ink in the

> arbitrary characters of a highly developed language
> the statement that civilization is the result of
> natural laws, and that man's duty is to let nature
> alone so that untrammeled it may work out a higher
> civilization, he simply ignores every circumstance
> of his existence and deliberately closes his eyes
> to every fact within the range of his faculties.
> If man had acted on his theory there would have
> been no civilization, and our philosopher would
> have remained a troglodyte.[7]

Ward proceeded to advocate a combination of democratic politics and social engineering, which he called "sociocracy" and which he saw as re-
pudiating both *laissez-faire* individualism and Marxian socialism.[8]

Another nineteenth-century debate of profound social importance
was also based on an appeal to scientific principles, in this case the
principle of the conservation of energy. The argument concerned the
issue of higher education for women. Spurious arguments, based on purported scientific facts, rationalized denying women access to higher
education.[9] The argument rested on the premise that in each individual
the principle of the conservation of energy governs the life forces.
Each person possesses a finite quantity of energy, and it may be employed for either reproduction or intellectual development, but there
is not enough for both, especially in women.[10] Some medical textbooks
of the day even contained illustrations depicting physical connections
between the gonads and the brain: the over-development of one system
was bound to lead to atrophy of the other.[11] According to this view,
the stages of sexual development in a woman's life, particularly men-
struation, call for vast amounts of energy. Since the brain draws energy from the rest of the body during intellectual activity, too much
energy devoted to studies, particularly during the critical developmental periods of adolescence and young adulthood, would cause the body to
suffer and fertility to diminish.[12] That this theory had practical
consequences is evident. For example, in 1877, the Board of Regents of
the University of Wisconsin, a coeducational institution, justified offering different curricula to men and women on the following grounds:

> Every physiologist is well aware that at stated
> times nature makes a great demand upon the ener-
> gies of early womanhood and at these times greater
> caution must be exercized lest injury be done...
> Education is greatly to be desired, but it is bet-
> ter that the future matrons of the state should be
> without university training than that it should be
> produced at the fearful expense of ruined health;
> better that the future mothers of the state should

> be robust, hearty, healthy women than that by over
> study, they entail upon their descendants the germs
> of disease.[13]

The rise of evolutionary theory did not alter these arguments.
According to the influential book by Geddes and Thompson, *The Evolution
of Sex* (1889),

> The differences [between the sexes] may be exagger-
> ated or lessened, but to obliterate them it would
> be necessary to have evolution all over again on a
> new basis. What was decided among the prehistoric
> Protozoa cannot be annulled by an Act of Parliament.[14]

Sexual differences, ascribed to differences between the cellular metab-
olism of males and females, were the basis of different kinds of education and social roles for men and women.

Such examples are not confined to the nineteenth century and can
be multiplied without end. The controversy over race and IQ rages today. In the 1920s, apparent correlations between race and the low IQ
of immigrants just off the boat at Ellis Island were used to support
the anti-immigration legislation which had such a devastating effect on
refugees from Nazism a decade later.[15] The most recent example of all,
perhaps, is that of sociobiology which hopes to eliminate all of the
humanities and social sciences and replace them by appropriate branches
of genetics and biology.[16]

Further examination of such attempts to justify social programs by
appeals to scientific facts and theories will lead to a fuller discussion of the role of values in science. A two-pronged argument will
emerge from this discussion. On the one hand, there are many respects
in which scientific discourse is laden with values. It is therefore
not surprising that one can justify ethical or social programs by appealing to science: the relevant values may be present already in the
body of science itself. On the other hand, if there are instances of
value-free science (or if the scientific material appealed to is not
laden with values relevant to the social issues at hand), the attempt
to justify social programs by appeal to scientific knowledge will result in the commission of the naturalistic fallacy. That is to say,
these examples are all instances of fallacious reasoning: either the
values expressed in the conclusion lurk as suppressed premises in the
body of the science, rendering the appeal circular; or, the value conclusion simply does not follow from the scientific knowledge presented
as its justification.

I will attempt to demonstrate these points in the sections that
follow.

I. SCIENCE AND VALUES

First, let us examine the role of values in scientific discourse.

A hidden premise of those who would justify their social programs on the basis of scientific facts is the assumption that science itself is value-free. Otherwise, on what basis could they claim that science provides an objective, independent justification of the moral views in the context of which such programs are conceived? The conclusion will be doubly damaging to those who would use science as an independent justification of their moral and social views. On the one hand, in almost every respect, scientific discourse is laden with values; on the other hand, to the limited extent that science can be regarded as value free, its results are neutral *vis-a-vis* social application.

In order to establish the role of values in science, it will first be necessary to provide a brief description of what I consider to be the structure of scientific discourse. This description is merely a sketch, and for the purposes of the present paper it could be modified in many respects without harming the central thesis that scientific discourse bears a heavy load of values.

Scientific activity always takes place within a framework of assumptions about the nature of the world and the kind of knowledge it is possible for us to acquire about that world. I call this set of assumptions a "conceptual framework" and see it as providing answers to certain fundamental questions:[17]

1. Of what kinds of entities does the world consist?
2. In what ways do these entities interact? That is, by what means does change occur?
3. What kind of knowledge can we acquire about this world? What methods are appropriate for the acquisition of such knowledge?

In other words, the conceptual framework provides us with the ultimate terms of explanation for our science and with its epistemological and methodological stance. Thus, for Aristotelian physics, the physical world consists of matter and form (in the Aristotelian sense of these words), change occurs by means of the four causes, and knowledge is acquired by the intuitive act of grasping essences followed by demonstrative reasoning as spelled out in the *Posterior Analytics*. By contrast, for Cartesian physics, the world consists of matter (extension), and all change takes place by means of motion and impact. Knowledge is acquired by a combination of deduction from first principles which are

known *a priori* and certain observed facts about the actual state of the
physical world. Further examples of conceptual frameworks abound.

Conceptual frameworks cannot be established or rejected on scientific grounds, for they provide the criteria for establishing whether
a given statement or method is true or scientific. They are simply assumed - although there are always complex social, intellectual, and
historical reasons why one or another (or several!) conceptual frameworks are accepted at any given time. Within any conceptual framework,
however, it is possible to formulate specific theories which are testable according to the methodological tenets set down in the framework.
Thus the development of various impetus theories in the middle ages can
be seen as an elaboration of the theory of motion, couched in the Aristotelian language of matter, form, and the four causes. These theories
were accepted or rejected according to Aristotelian criteria for determining essences. Similarly, the various 17th-century attempts to
derive the sine law of refraction - those of Descartes, Huygens, Fer-
mat, and others - can be understood as a search for a mechanical model
of refraction which were subject to scientific test, as defined by the
mechanical philosophers.

Finally, the conceptual framework and the particular theory couched within its terms conspire to determine what sorts of things will
constitute facts and evidence. To borrow and amplify an example from
Kuhn, the same physical system which Galileo saw as a pendulum, Aristotle would have seen as an instance of constrained natural motion, and
David would have seen as a weapon for slaying Goliath. What we see,
what constitutes a fact, when we regard the world, will be determined
by a complex network of assumptions.[18]

Given this analysis of the structure of science, let us now undertake an examination of the role of values in science. If the choice of
conceptual framework cannot be made on "scientific" grounds, how is it
made? It seems to me that such choices are made on pragmatic considerations: what do we wish to accomplish by means of our science? Do we
want a science that is part of a holistic world view, one which is in-
tegrated with theology and political philosophy? Or, do we want one
which will enable us to make precise mathematical predictions or be
useful for technological applications, quite apart from any relationship it might have to other areas of human activity? Such considerations lie behind choices of conceptual frameworks - as for example the
choice between the Aristotelian and Galilean approaches to physics in

the early 17th century. But such considerations are ultimately based
on values: which system is a better means to a stated end? Which ends
are the most compatible with our values? These values may be moral
values, especially in the social sciences where choice of conceptual
framework may depend on what one considers the just society to be or
what model of human nature best expresses one's moral outlook. Sometimes the values involved are aesthetic, as when a mathematician
chooses one set of axioms over another on the basis of elegance or econ-
omy. Often such choices are related to explicitly ideological concerns.
The Soviet scientists' rejection of the Copenhagen interpretation of
quantum mechanics and their insistence on a deterministic interpretation of physics was a result of their adherence to historical determinism which, in turn, depends on physical determinism.[19] But in every
case, the choice of framework boils down to value considerations.

The terms of explanation which the scientific community finds ultimate at any particular time - that is, the conceptual framework which
it chooses to adopt - reflect the values of the scientists, who are
themselves members of the community at large. A society in which race
were not an important social problem would not be much concerned with
the question of whether there is any relationship between race and IQ,
just as we would find it absurd to spend vast sums of research money on
determining whether there is any significant correlation between IQ and
curly hair. Fashions in science reflect changes in prevailing values
in the social milieu. The presently rising interest in the psychology
of women, evidenced by the proliferation of university courses and research on this subject, is surely a reflection of the questions raised
by the feminist movement. In the 19th century, the growing importance
of the various sciences in the curricula of the German universities can
be traced to an increasing awareness of the social or economic value of
these subjects.[20] In many respects, therefore, value considerations
enter into the ultimate framework of assumptions within which scientists operate. Although the choice of framework cannot always be related easily or directly to prevailing social and economic conditions,
nonetheless, values of one sort or another govern the choice of conceptual framework.

At the level of scientific theory, the position one takes in a
scientific debate may be the reflection of one's own value system.
Jonathan Harwood has shown that the two camps in the race-IQ controversy, the hereditarians and the environmentalists, can be sorted along

lines determined by the social origins and political opinions of the
scientists involved in the debate.[21] And Hugh Lacey argues in a recent paper that many theoretical debates in psychology result from ir-
reconcilable differences in views of human nature assumed by opposing
theorists. The views of human nature, in turn, rest on various kinds
of value assumptions.[22] A most blatant example of the domination of
scientific theory by value considerations occurred in the history of
psychology of sex differences, where characteristics thought to be associated with rationality were measured as consistently greater among
men than women, even if that assumption resulted in outright contradiction of earlier measurements. When the frontal lobe was thought to be
the seat of rationality, men were measured as having larger frontal
than parietal lobes. When theory changed and the rational function of
the parietal lobe increased, so did the relevant measurements.[23] Thus
even scientific "facts" are not insulated from value choices.

Another area in which values play a critical role is the choice
of applications of science. The choice to apply our scientific knowledge to produce vaccine or nerve gas, fertilizer or napalm, is clearly
goal-directed. And these goals, in turn, are the concrete expression
of our values. The underlying ethical issues are readily apparent in
many debates in contemporary medicine, e.g. whether limited medical resources should be allocated for heart transplants for a privileged few
or for large-scale development of preventative medicine.

Is there any aspect of science that is in fact value free? What
has become of the myth of the objectivity and disinterestedness of
scientific inquiry? If science is as thoroughly value-laden as the
present discussion suggests, any attempt to justify a moral stance by
appeal to scientific facts or theories would seem to run the risk of
circularity: either the value in question lurks as a suppressed premise in the scientific material - the case of the school of fish - or
the values in question are actually assumed as part of the conceptual
framework - the heredity-environment debate. Such circularity de-
stroys the possibility of finding an independent justification for our
values in science.

There are however two respects in which science may be considered
to be value-free, or in which the values assumed do not intrude upon
the issues at hand to render the moral argument circular. First, to a
limited extent, science can be regarded as free from the prejudices of
the individual investigator. Once a question is posed, within a

framework of assumptions, and once methods and procedures are agreed upon, the results of scientific investigation are value-free *only* in the sense that the outcome of the investigation is determined by the answers generated in the investigative process and not by the predilections of the scientist. Secondly, the values assumed as part of the framework of the science may have no bearing on whatever social or moral prescriptions we want to justify. Thus, the preference for one statistical theory over another on the basis of mathematical elegance may have no bearing at all on the question of race and IQ or the decision to educate members of one race differently from those of another. But this value-neutrality is limited in the extreme. The history of science teaches us that the choice of assumptions and of methods as well as the choice of questions to be investigated are choices based on values.[24] Granted the value-laden quality of the framework of the investigation, how do these limited areas of neutrality pertain to the larger issue of the use of science to justify moral positions?

Not at all. Precisely to the extent that the results of scientific inquiry are value-free, they are neutral with regard to social application. Consider studies in chemistry. Granted the fact that the general framework for chemical investigation may have been determined by some set of social needs or values, the placement of some element into a particular slot in the periodic table was a value-free decision: once having determined its atomic weight and atomic number, the placement followed in a perfectly objective way. But the kind of knowledge generated in this manner, though it may enable us to manufacture synthetic dyes and fertilizers, atomic bombs, napalm, or saccharine, in no way helps us to choose which we are to make, or, more generally, to answer the moral questions which lie at the heart of such choices.

Is there then no moral to be drawn from the school of fish?

II. THE NATURALISTIC FALLACY

The moral of the school of fish is that there is no moral to be drawn from their biological nature. Many examples of attempts to justify moral conclusions by appeal to natural facts illustrate the same point: opposing consequences can be justified on the basis of ostensibly identical premises.[25] There is something downright fishy about this mode of reasoning.

The problem boils down to what philosophers ever since Henry Sidgwick and G. E. Moore have called the "naturalistic fallacy". The problem was recognized earlier by Hume.

> I cannot forbear adding to these reasonings an
> observation, which may perhaps, be found of some
> importance. In every system of morality, which
> I have hitherto met with, I have always remarked
> that the author proceeds for some time in the or-
> dinary way of reasoning, and establishes the being
> of God, or makes observations concerning human af-
> fairs; when of a sudden I am surprised to find,
> that instead of the usual copulations of proposi-
> tions *is* and *is not*, I meet with no proposition
> that is not connected with an *ought* or *ought not*.
> This change is imperceptible, but is, however, of
> the last consequence. For as this *ought*, or *ought
> not*, expresses some new relation or affirmation,
> 'tis necessary that it shou'd be observ'd and
> explained; and at the same time that a reason
> should be given, for what seems altogether incon-
> ceivable, how this new relation can be a deduction
> from others, which are entirely different from it.[26]

Whether or not deriving an "ought" from an "is" is logically fallacious,
Hume argued that such an inference requires explanation and cannot remain unjustified. All attempts to justify ethical conclusions by deriving moral statements from scientific facts, of course, involve an
inference from an "is" to an "ought".

Sidgwick and Moore both wrote in the wake of the scientific naturalism that came to be so popular in the late 19th-century Britain.[27]
The scientific naturalists, T. H. Huxley, John Tyndall, Herbert Spencer,
and others of their ilk, were keen to show that all the characteristics
heretofore considered uniquely human could be reduced to scientifically
understandable biological facts, especially those of Darwin's theory of
evolution. Accordingly, human reason, spirituality, and ethics were to
be explained as the outcome of purely biological processes, just as today's sociobiologists want to demonstrate that altruism and other apparently uniquely human characteristics are the result of evolution understood through the principles of population genetics.[28]

Sidgwick's revulsion against the stark scientific world of the naturalists led him to turn to psychical research as well as to consider
the philosophical basis of ethics. Moore's profoundly influential
Principia Ethica, first published in 1903 and still widely studied in
the English-speaking world, opened with a strong attack on the kind of
reductionism advocated by the scientific naturalists. Coining the term
"naturalistic fallacy", Moore set out to argue that any attempt to

define "good" in terms of some natural property - such as "fit" in the
biological sense - is not possible. The primary target of Moore's attack was the evolutionary ethics of Herbert Spencer.[29]

Let us consider Moore's version of the naturalistic fallacy in
some detail, because it lies at the heart of the problem with the argument from the school of fish. Moore argued as follows. The naturalistic fallacy[30] consists in the assumption that because some quality,
such as "fit", always and invariably accompanies the quality of goodness that therefore this other quality is identical with goodness.
This assumption is analogous to arguing that because "octagon" and
"red" always denote the same stop-signs that octagonal and redness are
one and the same quality. If the word "good" and the word "fit" describe the same things but do not attribute the same quality to them,
then to say "what is fit is good" is to make a significant statement:
its negation is not a self-contradiction, and it tells us something we
cannot glean simply from examining the meanings of the words involved.
But if "good" and "fit" also have the same meaning, that is if they refer to identical qualities, then the assertion "what is fit is good" is
an empty tautology. On the basis of such considerations, Moore argued
that "good" cannot be defined, because statements of the form "X is
good" (where "X" is some potentially defining quality of "good") always
turn out to be significant, in the sense that the further question,
"But is X really good?" remains meaningful. It may be an empirical
fact that "good" is universally associated with some natural property
such as "biological fitness". But, Moore insisted, these associated
natural properties do not exhaust the meaning of "good", and therefore
the reduction of ethics to some branch of natural science is not possible. "Good", like "yellow", is a basic term of our vocabulary, and
cannot be eliminated by substituting for it some other term which fully
captures its meaning.[31]

Moore proceeded to examine specific ethical theories in which the
naturalistic fallacy has been committed, and he showed just how the
fallacy caused them to break down. He selected Spencer's evolutionary
ethics as a particular target and showed that any attempt to define
"good" in a naturalistic sense would reduce ethics to an empirical science. Such a reduction would result in the elimination of ethics as an
autonomous subject.

> I have thus appropriated the name Naturalism to a
> particular method of approaching Ethics - a method
> which, strictly understood, is inconsistent with
> the possibility of any ethics whatsoever. This
> method consists of substituting for "good" some
> one property of a natural object or a collection
> of natural objects; in thus replacing Ethics by
> some one of the natural sciences.[32]

But when he applied his analysis of the naturalistic fallacy to the
specific case of Spencer's theory, such a reduction of ethics to an empirical science did not succeed in eliminating the underlying ethical
question.

> The survival of the fittest does *not* mean, as one
> might suppose, the survival of what is fittest to
> fulfill a good purpose - but adapted to a good end:
> at the least, it means merely the survival of the
> fittest to survive; and the value of the scientific
> theory, and it is a theory of great value, just con-
> sists in shewing what are the causes which produce
> certain biological effects. Whether these effects
> are good or bad, it cannot pretend to judge.[33]

Indeed, the only way in which one could derive ethical conclusions from
the biological principle of the survival of the fittest would be by im-
parting ethical content to the purely biological notion of "fitness".
While Spencer still held the view that humanity stands at the top of
the evolutionary tree, a more sophisticated biologist (such as Darwin
himself) might argue that mosquitoes are just as fit as men, given the
different ecological niche each species occupies; for biological fitness is always relative to a particular environment. Indeed the bac-
teria that live in the human gut might even be considered more fit than
their human hosts, given the fact that as parasites they have had to
adapt to the human context. What then becomes of the seemingly simple
statement that the fit is the good?

Two conclusions emerge from this discussion. The first is that in
any attempt to derive an "ought" from an "is", either the attempt fails
outright, or the "is"-statement already contains a hidden "ought"-
statement.[34] Various examples considered in this paper illustrate that
point. Spencer's version of social Darwinism contains the hidden premise that evolution is morally progressive and that the principle of the
survival of the fittest operates directly on individuals. These assumptions, already hidden in the science to which he appeals to "justify"
his doctrine of *laissez-faire* individualism, result in question-begging.
Similar considerations apply to Ward's advocacy of social engineering.
The story of the school of fish is paradigmatic of these cases. The

facts of their biological nature do not lead us to any conclusions what-
soever about human nature or the status of women. Obviously, for both
the feminist and the male chauvinist, either the scientific facts are
entirely irrelevant to their ideological stances, or each has inserted
an unstated premise into the biological data, a premise which contains
the values relevant to their social views.

The second conclusion follows hard upon the first. Given the vicious characteristic of these naturalistic systems of ethics, namely
the fact that several incompatible conclusions apparently follow from
the same set of premises, they utterly fail to provide a scientific basis for ethics. Indeed, supposing that Spencer successfully justified
his social Darwinism by appealing to biological principles and supposing that Ward was equally successful in justifying his theory of social
engineering by appeal to the very same biological principles, the fundamental ethical question, "what is the good society?" or "what approach to social problems is the best?" remains unanswered. In other
words, the fundamental question of ethics, far from being answered by
appeal to scientific facts, remains as elusive as ever. Crudely put,
feminists will find the school of fish just as useful to their ends as
will male chauvinists. And insofar as both are guilty of using bad
arguments, both are subject to the same kind of criticism. To overcome
this problem, each group will have to come to grips with its own values
in an explicit way.

III. CONCLUSIONS

I am not proposing to develop a theory of ethics in this paper. I
am, however, arguing that the problems of ethics cannot be avoided by
appealing to scientific facts. Perhaps, as some philosophers have argued, values are just as fundamental as facts, or even more so.[35] Perhaps there is no clear distinction between facts and values, as recent
analyses of the structure of science might suggest.[36]

The point I am determined to make is that we must come to grips
with our values when making social prescriptions. Since we do so any-
way, we might as well do so consciously. Scientific knowledge will not
do this for us. Science may inform us of some of our possibilities and
some of our limitations.[37] It will not tell us what we ought to do
within these constraints. Supposing we discover, by scientific investigation, that there is indeed a fifteen-point disparity between the

scores of whites and blacks on IQ tests. What does this tell us to do?
Should we prescribe different kinds of education for members of each
race? Or should we sort individuals according to their own results on
the tests? To answer these questions, we must decide whether to treat
people as individuals or as members of groups, and this decision depends on whether our value system is based on individual or collective
responsibility. Likewise the results of chemical experiments will not
tell us whether to manufacture napalm or fertilizer or both. That decision rests on the goals and values we hold or to which we aspire.
There are certainly no built-in scientific reasons why we must approach
these problems one way or another.

Even if scientific knowledge cannot provide us with solutions to
our moral dilemmas, it is nonetheless not entirely irrelevant to such
concerns. Science can tell us what is possible; it can spell out our
physical and physiological limitations; it enables us to project the
consequences of various courses of action. Knowledge of these things
is directly relevant to moral choices, but it does not provide answers
to moral questions. The moral problem remains when we come down to the
matter of deciding which of several courses of action to adopt.

Science will tell us many things about what we can and cannot do;
it cannot, however, free us from the fundamental human task of making
value choices.

FOOTNOTES

1   My special thanks are due to J. J. MacIntosh, who first told me
    about the school of fish and inspired me to write this article. I
    have subsequently learned that these fish do indeed exist and are
    called Wrass. Thanks are also due to several colleagues and friends
    who have read drafts of this paper and have helped me clarify my
    ideas through numerous discussions: Marsha Hanen, Eliane L. Silver-
    man, Sheldon A. Silverman, Martin Staum, Petra von Morstein, and
    Robert G. Weyant. Earlier versions of this paper were presented to
    the Sigma Xi Chapter at the University of Calgary in January 1978,
    to the Department of Philosophy at the University of Lethbridge in
    March 1978, and to the Department of General Science at Oregon State
    University in June 1978.

2   The most influential of Spencer's many books was *Social Statics; or,
    the Conditions Essential to Human Happiness Specified, and the First
    of Them Developed*. Although first published in 1851, *Social Statics*
    did not have great impact until after the publication of Darwin's
    Origin of Species in 1859. Spencer also wrote a work which expli-
    citly attempted to base morality on scientific foundations to re-
    place what he considered an obsolete morality based on supernatural
    foundations. See The *Data of Ethics* (1879). Spencer's intellectual
    development is more complex than that depicted in most traditional
    accounts. See Mark Francis, "Herbert Spencer and the Myth of
    Laissez-faire," *Journal of the History of Ideas*, 1978, 39: 317-328.

3   Richard Hofstadter, *Social Darwinism in American Thought*, revised
    edition (Boston: Beacon Press, 1955), pp. 35-41. For further de-
    tails on the tenets and dissemination of Social Darwinism, see:
    Robert C. Bannister, "'The Survival of the Fittest is our Doctrine':
    History or Histrionics," *Journal of the History of Ideas*, 1970, 31:
    377-398; R. J. Halliday, "Social Darwinism: A Definition," *Victori-
    an Studies*, 1971, 14: 389-405; Sidney Ratner, "Evolution and the
    Scientific Spirit in America," *Philosophy of Science*, 1936, 3:   104-
    122; James Allen Rogers, "Darwinism and Social Darwinism," *Journal
    of the History of Ideas*, 1972, 33: 265-280; R. M. Young, "The Im-
    pact of Darwin in Conventional Thought," in A. Symondson, ed., *The
    Victorian Crisis of Faith* (1970).

4   William Graham Sumner, *The Challenge of Facts and Other Essays* (New
    Haven: Yale University Press, 1914), p. 90, quoted by Hofstadter,
    op. cit., p. 58.

5   Quoted by Clifford H. Scott, *Lester Frank Ward* (Boston: Twayne Pub-
    lishers, 1976), p. 91. Ward's views are also discussed by Hof-
    stadter, *op. cit.*, Chapter 4.

6   Lester Ward, *Dynamic Sociology* (1882), quoted in Henry Steele Com-
    mager, ed., *Lester Ward and the Welfare State*, p. 50.

7   Lester Ward, "Art is the Antithesis of Nature," *Mind*, 1884, 9,
    quoted by Commager, *op. cit.*, pp. 78-79.

8   Scott, *op. cit.*, p. 86.

9   This topic is the subject of much recent research. See: Flavia
    Alaya, "Victorian Science and the 'Genius' of Women," *Journal of
    the History of Ideas*, 1977, 38: 261-280; Carol Dyhouse, "Social
    Darwinistic Ideas and the Development of Women's Education in
    England, 1880-1920," *History of Education*, 1976, 5: 41-58; Eliza-
    beth Fee, "Science and the Woman Problem: Historical Perspectives,"
    in Michael S. Teitelbaum, ed., *Sex Differences: Social and Biolo-
    gical Perspectives* (New York: 1976); Janice Trecker, "Sex, Sci-
    ence, and Education," *American Quarterly*, 1974, 26: 352-366; Ben
    Barker-Benfield, "The Spermatic Economy: A Nineteenth-Century View
    of Sexuality," *Feminist Studies*, Summer 1972, 1: 45-74; Vern Bul-
    lough and Martha Voght, "Woman, Menstruation, and Nineteenth-Cen-
    tury Medicine," *Bulletin for the History of Medicine*, 1973, 47:
    66-82; Joan N. Burstyn, "Education and Sex: The Medical Case
    Against Higher Education for Women in England, 1870-1900," *Proceed-
    ings of the American Philosophical Society*, 1973, 117: 79-89;
    Elaine Showalter and English Showalter, "Victorian Women and Men-
    struation," in Martha Vicinus, ed., *Suffer and Be Still: Women in
    the Victorian Age* (Bloomington: Indiana University Press, 1972);
    Carroll Smith-Rosenberg, "Puberty to Menopause: The Cycle of Femi-
    ninity in Nineteenth-century America," in Mary S. Hartman and Lois
    Banner, eds., *Clio's Consciousness Raised: New Perspectives on the
    History of Women* (New York: Harper and Row, 1974); Carroll Smith-
    Rosenberg and Charles Rosenberg, "The Female Animal: Medical and
    Biological Views of Woman and Her Role in Nineteenth-century Ameri-
    ca," *Journal of American History*, 1973, 60: 332-356; Martha Ver-
    brugge, "Women and Medicine in Nineteenth-century America," *Signs*,
    1976, 1: 957-972; Ann Wood, "The Fashionable Diseases: Woman's
    Complaints and Their Treatment in Nineteenth-Century America,"
    Journal of Interdisciplinary History, 1973, 4: 25-57.

10 Jill Conway, "Stereotypes of Femininity in a Theory of Sexual Evo-
    lution," *Victorian Studies*, 1970, 14: 48.

11 Carroll Smith-Rosenberg and Charles Rosenberg, *op. cit.*, 334-341;
    Elaine Showalter and English Showalter, *op. cit.*, 86. This same
    theory was used to account for the fact that Negroes were supposed-
    ly stupid and over-sexed. According to the then-popular theory,
    members of that race reached sexual maturity earlier than whites
    with the result that energies which might otherwise have been used
    for intellectual development were channelled to the sexual organs
    instead. See John S. Haller, Jr., *Outcasts From Evolution: Scien-
    tific Attitudes of Racial Inferiority, 1859-1900* (New York: McGraw-
    Hill Book Co., 1971), p. 52.

12 Showalter and Showalter, *op. cit.*, pp. 85-88.

13 Quoted by Smith-Rosenberg and Rosenberg, *op. cit.*, pp. 341-342.

14 Quoted by Conway, *op. cit.*, p. 53.

15 Leon J. Kamin, *The Science and Politics of IQ* (New York: John
    Wiley and Sons, 1974), Chapter 2; Kenneth M. Ludmerer, "Genetics,
    Eugenics, and the Immigration Restriction Act of 1924," *Bulletin of
    the History of Medicine*, 1972, 46: 59-81.

16 "In this macroscopic view the humanities and social sciences shrink to specialized branches of biology; history, biography, and fiction are the research protocols of human ethology; and anthropology and sociology together constitute the sociobiology of a single, primate species." Edward O. Wilson, *Sociobiology: The New Synthesis* (Harvard University Press, 1975), p. 547.

17 My analysis here owes a lot to R. Harré, *Matter and Method* (London: Macmillan and Co. Ltd., 1964).

18 See for example, Feyerabend's long discussion of what was involved in Galileo's telescopic observation of the satellites of Jupiter in *Against Method* (London: Verso, 1975), Chapters 10 and 11.

19 Loren R. Graham, *Science and Philosophy in the Soviet Union* (New York: Alfred A. Knopf, 1972), Chapter III.

20 Joseph Ben-David, *The Scientist's Role in Society: A Comparative Study* (Englewood Cliffs, New Jersey: Prentice-Hall, Inc., 1971), Chapter 7; Frank Pfetsch, "Scientific Organization and Science Policy in Imperial Germany, 1871-1914: The Foundation of the Imperial Institute of Physics and Technology," *Minerva*, 1970, 8: 557-580.

21 Jonathan Harwood, "The Race-Intelligence Controversy: A Sociological Approach," *Social Studies of Science*, 1976, 6: 369-394; 1977, 7: 1-30. A similar ideological alignment among German physicists during the Weimar period has been noted by Paul Forman, "The Financial Support and Political Alignment of Physicists in Weimar Germany," *Minerva*, 1974, 12: 39-66.

22 Hugh M. Lacey, "On the Value Neutrality of Psychological Inquiry," unpublished manuscript.

23 Stephanie A. Shields, "Functionalism, Darwinism, and the Psychology of Women: A Study in Social Myth," *American Psychologist*, 1975, 30: 739-754. On the question of physical and intellectual differences among the human races, similar finagling of data on ideological grounds has occurred. See Stephen Jay Gould, "Morton's Ranking of Races by Cranial Capacity," *Science*, 1978, 200: 503-509.

24 It is a well known historical fact that one of Copernicus' primary motives for reforming Ptolemaic astronomy was his desire for a more harmonious and unified system of astronomy. He did not base his revolution on new observations but rather on a metaphysical preference for certain forms of mathematical harmony. See Thomas S. Kuhn, *The Copernican Revolution* (Cambridge, Massachusetts: Harvard University Press, 1957). For Copernicus' statement of his motives, see Nicolaus Copernicus, *On the Revolutions of the Heavenly Spheres*, trans. Charles Glenn Wallis (Chicago: Encyclopedia Britannica, Inc.), pp. 507-508.

25 For a penetrating and delightful critique of some modern, popular attempts to reintroduce this kind of biological determinism, see Stephen Jay Gould, *Ever Since Darwin: Reflections in Natural History* (New York: W. W. Norton and Co., Inc., 1977), esp. Chapters 31 and 32.

26 David Hume, *A Treatise of Human Nature* (London, 1739), p. 469.

27 Frank Miller Turner, *Between Science and Religion: The Reaction to Scientific Naturalism in Late Victorian England* (New Haven: Yale University Press, 1974), Chapters 1 and 2.

28 The final chapter of Wilson, *Sociobiology* makes this goal plain. For a critique of sociobiology compatible with the argument in this paper, see Gould, *Ever Since Darwin*, Chapters 32 and 33.

29 George Edward Moore, *Principia Ethica* (Cambridge: Cambridge University Press, 1903), Chapter II. Much recent philosophical discussion has been devoted to the naturalistic fallacy. Most of it consists, on the one hand, of attempts to derive "ought" from "is," and, on the other, of attempts to refute such attempts. A collection of recent articles is contained in W. D. Hudson, ed., *The Is-Ought Question* (London: Macmillan, 1969). See also Philippa Foot, ed., *Theories of Ethics* (Oxford: Oxford University Press, 1967).

30 My account of the naturalist fallacy in the following paragraph is a paraphrase of Arthur N. Prior, *Logic and the Basis of Ethics* (Oxford, 1949), pp. 1-8.

31 Moore, *op. cit.*, p. 10.

32 *Ibid.*, p. 40. The validity of Moore's claim that naturalism would lead to the replacing of ethics by one of the natural sciences is borne out by the stated aim of Wilson's *Sociobiology*. See footnote 16 above.

33 Moore, *op. cit.*, p. 18.

34 The presence of the suppressed "ought" is illustrated in modern discussions by the much discussed paper by John R. Searle, "How to derive 'Ought' from 'Is,'" *Philosophical Review*, 1964, 73: 43-58 (reprinted in Foot, *op. cit.*).

35 Lacey, *op. cit.*

36 Feyerabend, *op. cit.*

37 Gould, *Ever Since Darwin*, Chapters 30 and 32.

CLOSING REMARKS

In these closing remarks, I will attempt neither to summarize, nor to serve as a rapporteur (or raconteur) of the symposium. Rather, I would like to assess, after the fact, how far we have gotten with respect to the theme which the symposium title proposed. In making my opening remarks, I started out with a sort of hypothesis: I antici- pated what the symposium might be about by projecting from the title the sorts of issues and problems which the title suggested. The ques- tion now is: Were these in fact the issues and problems which were discussed? Did we come to any resolution of these questions, or make any progress concerning them? I have now had a chance to test the predictions which I made at the outset. I am batting about 230. (Good enough for some farm team perhaps, but not for the majors.)

Nevertheless, the papers and the discussion here revealed to me some most interesting features of the present stage of the interplay between the philosophy of science and the history of science. I am especially interested in this dialectic, which has some of the features of a courting dance and some of the features of a wrestling match. The game of "philosophy of science vs. history of science" began some years ago. In fact, in a recent article,[1] I trace it to the late nineteenth century, when the relations between the two fields were what I charac- terized as "Agreeable *external* relations". These then degenerated into "Disagreeable external relations", progressed to "Agreeable" and then "Disagreeable *internal* relations", and we now stand at the threshold of a new stage of "Agreeable internal relations". Neither side has "won" the game, and neither has yet been fully *aufgehoben* into some higher synthesis. But things are certainly different than they used to be. I remember Russell Hanson as saying, at a conference in about 1962 or '63, that there were *no* relations between history and philosophy of science, and that any attempt to establish such relations should be regarded as so much "waffle". (This was especially funny coming from Hanson, but that's how things used to be.) At our present stage, the interaction between the two fields is lively though still problematic. Let me characterize how I have perceived it here, in our consideration of the question of the relations among science, pseudo-science and society.

First of all, the title was, I still think, very well chosen. The
three terms of the title did generate a kind of an interplay among the
concepts. The science/pseudo-science issue raised the old demarcation
problem in a new way, because in the older discussion, what was demarcated from science was not called pseudo-science. It wasn't then a
question of marking off science from pseudo-science, but rather of
distinguishing science from things which were not properly scientific,
like metaphysics, poetry, or faith. "Pseudo-science" is a much tough-
er, much more critical characterization than that which the traditional
demarcation problem proposed. And then, adding "society" in the title
as a context for our considerations added all the jazzy externalist
stuff that has sparked the discussion of the last ten to fifteen years.
It is no longer a clean-cut difference in methodology between histori-
cist externalists on the one hand and, on the other, internalists and
formalist philosophers of science who are ahistorical. Rather, we
have been mucking around and trying very hard to see what the more integral relationships are between social context and the development of
scientific theory, and what the criteria are for what counts as science, not simply in the course of its social practice, but in the
course of its historical development. So I think that what I have
sensed here is certainly a much more philosophically informed, historical discussion and a much less recalcitrant mood on the part of the
philosophers of science to take the historical context, or the historical facts seriously.

However, if I may divide the particular sorts of questions which
came up and were discussed in the various papers, into the philosophy
of science and the history of science, for purposes of analysis, then
a distinctive emphasis appears in the historical papers (if one can
call it an emphasis, since it is a *negative* emphasis). There appears
to me to be an overwhelming attempt, among the historians, to evade
normative questions, or the resolution of normative issues which arise
historically; and in particular, the normative issue concerning the
demarcation between science and pseudo-science. This evasion is not
naive, but apparently deliberate. It is not as if the historians are
not aware that the normative issue is there, but rather as if there
were a conscious division of labor. It is as if the historians were
trying to say: "Look here. We know there is a question concerning
what is properly scientific and what is not, and that there is a need
for historical judgments as well as contemporary judgments of this

sort. But, as historians, we are not trying to solve these questions. Rather, we can do no more than to provide some of the materials which may go into a resolution of these questions." There is, if I may say so, a certain kind of positivism in the very historicity of the historians here, in refusing to take on the normative questions full tilt. Now I should add that this objectivist sort of stance was given up by our two younger-generation historians (Cooter and Osler). They took on the value questions head on, or feet first. But the more cautious approach of a majority of the historical papers was one which I think feigned a certain dispassionate approach toward making normative determinations as to who was right, and whether it was "good for the Jews", that is to say, whether it helped scientific progress or not, or whether it would have been better to do it some other way, or whatever. Thus, the historical papers on the whole seemed to presuppose one sort of resolution of the question of the conference, without actually stating this resolution. The question posed in the conference was how (and perhaps whether) to make a distinction between science and pseudo-science. But in a number of historical papers, the presupposition seemed to be that pseudo-science is whatever was taken to be questionable as science in its own time. In these papers, no normative judgments were made on this question, but rather it appeared that we were simply getting a report on what the norms happened to be. Now it is certainly a part of historical analysis to reconstruct the *facts* about normative judgments, as they figured in history. Contemporary attitudes are part of what constitutes the historical reality, and the explanation of why such attitudes were taken, or why certain judgments were made helps us to understand history. But one might demand a further philosophical judgment, even from historians of science, as to what did and what did not constitute science proper, at various stages of the historical development of science. *Some* theory of science operates, even tacitly, in every history of science, and it would be, in my view, a philosophical advance if historians of science were able to articulate their own theories of science, as part of their analytical and methodological repertoire. Thus, for example, Sam Westfall makes it clear, in his paper, that alchemy was taken to be questionable, as science proper, in Newton's time. The evidence is that Newton, despite the fact that he did so much with alchemy, and wrote so much on it, did so with a certain shyness, and a certain recognition that it didn't have the same status as his other work, say in mechanics or optics. The impression

I got, from Westfall's paper, was that Newton himself recognized that alchemy was not quite science proper, though he took it very seriously. This gives us a vivid sense of the complexity and agony of Newton's own philosophical and scientific thought. But it does not yet tell us where *Westfall* stands on the question of the relation of alchemy to science proper, in the historical development of Newton's thought. Was alchemy pseudo-science then? Was the contemporary attitude (or Newton's own attitude) correct, *in its time*? Similarly, in the Jacobs' paper, it is shown in a striking way that, by the 1650s, the Puritan reformers were ready to turn away from the earlier acceptance of alchemy, astrology and number mysticism, on political grounds; for these "sciences" were associated with the more radical and enthusiastic elements of the movement, who were now being distanced by a more conservative Puritan politics. This paper was particularly interesting in that it gave a very clear sense of what leading intellectual and political figures at that time thought was legitimate in science, and what not; and it also gave a clear sense of what the political issues were which affected these normative judgments about science and pseudo-science. Nevertheless, here too, one could only tease out what the standpoint of the historian was, with respect to what was to be taken appropriately as science proper, in that context. We are being presented, then, with a kind of value-free history, at least with respect to the science/pseudo-science issue. There is, in the Jacobs' paper, an implicit political judgment, that the conservative shift in Puritan politics was a Bad Thing. There is also the useful explanation of how politics influenced judgments about science. But there is, as yet, no normative judgment about the question of the status of the alleged pseudo-sciences. Were they *pseudo* or not, in their own time? Were the Puritan reformers scientifically right, even if politically wrong? Or is history "just the facts", and the historian their faithful recorder and revealer?

This is all the more important, since the context is that of Newtonian science, and not pre-Newtonian or post-Newtonian. And it is this Newtonian science of the seventeenth century that we generally take to be the paradigm (you should pardon the expression) of modern science. Now we take this in a critical, not an uncritical sense, and we don't simply say that science proper is identical with Newtonian science. But still, this character of Newtonian science as the paradigmatic form of modern science is generally accepted. And therefore,

it becomes a crucial matter to determine what the historical relation
was between all these other "sciences" - alchemy, astrology, number-
mysticism - and Newtonian science. To what extent did these other
sciences contribute to, hamper, intrude upon, the development of modern
science? Were they digressions, vestiges or constitutive aspects of
modern science? This is a *normative* historical question, and thus the
proper subject for an informed historical and philosophical judgment.
Yet, the historians, it seems to me, largely stay on their side of the
disciplinary fence here, and I am inviting them to tear the fence down.

Another interesting feature of the historical papers, as well as
of the philosophical ones, is that the sciences taken up as prospectively pseudo-scientific fall into two groups: the first, made up of *traditional* sciences (or pseudo-sciences, or protosciences), like astrology and alchemy; the second, made up of *new* sciences (or pseudo-sciences
or protosciences). These latter are the ones Flew alluded to as "as-
pirant" sciences, in his paper on Parapsychology. To call any of these
protosciences is already to make a normative judgment, of course. It
suggests that these are on the way to becoming sciences proper, or are
the early stages of the development of a new science. What was striking about a number of the papers which dealt with such "aspirants" as
phrenology, mesmerism or parapsychology, is that the reception and contemporary judgment of these "sciences", and in general, the judgment as
to what was, and what was not properly scientific was shown to be, at
the same time, an ideological and political judgment, and to be explained in part as a function of the social and political contexts of
the time. That's certainly normative in content, and the history, in
this sense, becomes a *history of norms*, though not yet a *normative
history*. One might characterize it as a descriptive normativism, in
the history of science, in that it gives an account of what were historically prevalent norms, without as yet making normative judgments
about these norms themselves. Such a history of norms is an essential
component of a normative, i.e., a critical history of science, without
which normative history would become either a matter of the imposition
of abstractly conceived norms upon the history of science, or a matter
of writing uninformed and tendentious histories. Thus, the history of
norms, even as a pre-critical history, is exceptionally worthwhile,
for as the papers at this conference showed, what was at issue was not
simply a characterization of political or ideological norms, but also
of methodological norms. And these methodological norms could then be

seen in their relation to the particular political and social contexts
of their times. This seems to me to be a major step forward in the
historiography of science - that is, to have achieved the degree of
sophistication in treating these questions, where something as techni-
cal and formal as methodology itself can be seen in its historical
development, and against the background of the social-political contexts in which such methodologies develop. Of course, this sort of
treatment neither supports a methodology as being right, nor does it
give us any grounds for criticizing it as wrong. That sort of *internal*
normative critique of methodology cannot be resolved by context. But
we can no longer continue to do the kind of context-free normative analysis of what is and what is not scientifically proper, or methodologically valid, without now taking into account the richness of context in
which methodology itself develops. Once the history of science becomes
also a history of scientific methods, i.e. a history of what was taken
to be the appropriate way of pursuing or achieving scientific knowledge, then such history cannot help but be a normative history, or
indeed, a philosophical history of science. For claims as to how science is to be done, properly speaking, amount to claims as to what is
proper science.

For example: In Levere's paper on Coleridge, it was made clear
that Coleridge's methodological approach to the critique of phrenology
and mesmerism, and the grounds he proposed for the acceptance or rejection of these purported sciences, were based not only on a particular
metaphysical or theological heuristic; but in addition, his methodology was also, in part, influenced by his politics, and his being an
Englishman. The fact of this influence does not, of course, make his
methodology better or worse. But it does permit us to view it in a
much richer context that that which considers only whether the methodology is logically consistent, or whether the arguments for organicism,
or against mechanism were good arguments. These latter considerations
of methodology, in terms of formal or logical criteria, are certainly
normative, and the sort we would expect of a good philosophical critique of Coleridge's thought. But this leaves out a major dimension:
the concepts "organicism" and "mechanism" are more than flat, timeless
conceptual alternatives. The concepts derive their richness, and
their concrete methodological content from the variety of human and
historical contexts and practices within which they developed. The
mechanism, or "mechanical philosophy" of the seventeenth century is

very different in spirit and in methodological application from four-
teenth-century mechanism, say, at Oxford or Paris; and both of these
differ from the kind of mechanism we are familiar with in contemporary
biochemistry or molecular genetics. Putting these concepts into their
distinctive historical contexts allows us a much richer methodological
understanding, and indeed, widens the framework of normative judgment
about such methodological concepts. Indeed, it permits us to take a
less parochial view of the ways in which such concepts as mechanism
and organicism are to be understood and assessed in contemporary science. In this sense, the historian's contribution, in reconstructing
the history of methodologies in science, is crucial to the task of
contemporary philosophy of science as well.

More generally, the question put to us by our literate CBC Inter-
viewer, "What is Science?" poses the same basic problem of the historical understanding of science. "What is Science?" is not simply a question about the present. Nor can one answer it in a flat etymological
way, or by appeal to the historical introduction of the term "scientist" in the nineteenth century, for the first time. In seeking to
establish what is continuous in the practice of science, we look for
those features of method, or content, which are common to seventeenth-
century science, nineteenth-century science, contemporary science. We
tend to establish *a priori* norms, or criterial features which will then
help us to pick out what we take to be properly scientific, in some
transhistorical, or even essential way. For example, we propose some
concept of rationality, or experimental test, or fruitful conjecture,
or explanatory or predictive power, or heuristic force as the hallmark
of science proper, at any time. But such an essentialistic normative
approach tends to overlook the very historicity of the concept of science, and also of the social institution of science. We should ask,
instead, what has *become* science, how did the very notion of science
emerge, or develop, up to our own time? We are no longer to be per-
mitted the virginal ahistorical naiveté of two generations ago, when
science was to be defined as what scientists did. There is some justice in that approach, of course. But which scientists were to be
taken as models? Presumably the great ones. How do we choose these?
By reputation. By achievement. That is to say, by the social certi-
fications of greatness. Or again, if not by personal model, then by
the criterion of what is accepted as scientific knowledge by the community of scientists, in an impersonal or transpersonal way. The

object of logical reconstruction of science is, ostensibly, the "body
of scientific knowledge." But the two-dimensional caricature of science which emerged from such an approach, useful as it may have been in
abstracting certain features of science, failed to capture the more
ragged dynamic of science as a human project, as a less-than-coherent
complex of modes of inquiry, of experimental ingenuity, of imaginative
speculation. To pick out the paradigmatic cases of science proper as
those upon which to base a reconstruction of what science *is*, properly
speaking, involves us in the *Meno* problem: We have to know what we
are looking for in order to know when we have found it.

Lakatos' enterprise was to pick out, first, who the great scientists were, and then to figure out what they were doing which could
provide the norm for good science. We start with Newton, Maxwell,
Einstein. (Perhaps also Marx; Lakatos wasn't sure about that; *not*
Freud, he *was* sure about that!) Add a few others. Then, if you can
figure out what they were doing, you would have a reconstruction of
what *real* science is. That gets you out of the *Meno* "circle" (or the
"hermeneutical circle") by a bootstrap operation. The alternative
would be to start with an *a priori* norm of *science proper*, and then
use it to select, as *scientists proper*, those whose work fitted this
norm. Newton, for example, would *have* to fit, or we would regard the
norm as wrong-headed. The trouble with this, however, is that such an
a priori approach would tend to lead us (and *has* led us) to look for
only those aspects of a scientist's work (say, Newton's) which fitted
the norm, and to disregard or discount other aspects. (This is the
well-known method in science of ignoring negative evidence.) But then,
such a method would be, effectively, self-certifying. If, alternatively, one is not going to go about answering the question "What is Science?" in such a simple *a priori* way, but take it in its historical and
concrete contexts as well, then "science proper" and "scientist proper"
will have to be assessed as *changing* concepts, and as contestable concepts, in a way which cannot ignore the difficulties.

In view of the historical sophistication which the papers offered
at this conference provide, we can no longer approach this question in
a naive way. Science, as the papers make clear, did not spring full-
grown from the head of Zeus. It grew, changed, remains an extraordi-
narily complex phenomenon. To add to the complexity, what the historians reconstruct, as the science of a certain time, *also* changes.
From Levere's paper, one may conclude that every age has its "Newton".

That is to say, the nineteenth century's understanding of Newton is
very different from our own. And in our own time, for example, Koyré's
Newton is different from Hessen's Newton, and both of these, from
McMullin's Newton. And clearly, Stillman Drake's Galileo is not
Husserl's, and neither of them bear much resemblance to Feyerabend's.
It may, of course, be the case that Newton is a more complex figure,
and Newtonianism a more complex phenomenon than any particular historian has been able to reconstruct. But then, this means that *if*
Newton is to be taken as the test of what some normative conception of
science takes to be science proper - i.e. if its adequacy is tested by
whether or not it captures the complex reality of Newton's science, -
then we cannot make do with anything less than the many-sided, con-
textually-rich historical reconstruction of Newton.

What I am suggesting here is that many of the historical papers
presented at this conference, which were ostensibly descriptive, were
in fact tacitly normative, despite the shyness of the approach to
normative judgments. The tacit normative component revealed itself insofar as we were able to tell, from the papers, who the "Good Guys"
were. We knew, for example, that Stubbe was coming out ahead, in
Jacob's paper. It wasn't simply a matter of "poor old Stubbe", and
how he has been misunderstood in his debates with the Royal Society.
Rather, it turns out Stubbe was a good radical, he was misinterpreted,
and he is being saved here by a proper interpretation. He's a Good
Guy after all. In Westfall's paper, we are not presented with Newton's
alchemical preoccupations as a sign of his *unscientific* soft-headed-
ness. We don't come away saying "My God! Even Newton was taken in by
that junk!" Rather, we are forced to think through what role these
preoccupations played in his scientific thought, and whether or not the
alchemical writings were a constitutive element in the development of
his thought. McKillop, in his paper, plainly tells us who the "Good
Guys" were: the "backward" Canadians. Thus, despite the ostensibly
non-normative cast of the historical papers, there was a large tacitly
normative component in many of them.

As to the straightforwardly philosophical-analytical papers, which
were expressly normative and critical on the issue of science and
pseudo-science, e.g. Thagard's paper on astrology, Grünbaum's, on the
critique of psychoanalytic "explanations", Flew's, on parapsychology:
I believe that all of them exhibit one common *negative* feature. That
is, there is something none of them do. None of them are into the

older, purely formalist game of doing what I might characterize as
"Third generation" logical reconstruction of science. Now, "first
generation" logical reconstruction of science was an exciting business. It was full of social passion, historical import, concrete reference to the science of its time; it was self-consciously political
as a philosophy of enlightenment opposed to dark reaction, dogma, blind
superstition, empty speculation. I'm talking about the Vienna Circle
and its immediate contexts. Any attempt to view these philosophers as
formalists simply fails to understand their project in *its* historical
context, and doesn't see what they were doing, or what they thought
they were doing. By the second and third generations, logical empiricism had become a fairly bloodless affair, and by the time it became
a small cottage industry in Anglo-American philosophy of science, it
could rightly be called formalistic. But no one here was doing that,
and fewer and fewer philosophers continue to do that, perhaps simply
because it has become boring. The philosophical papers offered here,
by contrast, were dealing with very concrete, even practical questions
in contemporary scientific thought. The question of what is pseudo-
science is not simply a conceptual question, nor will it yield to simply formal criteria. If some prospectively scientific endeavor fails,
as science, then to point out the respects in which it is failing, as
I think Grünbaum and Flew try to do, is a practical, applied question
in philosophy of science, which requires attention to the working hypotheses, the modes of validation, the actual practices and specific
arguments which the candidate for scientific status offers. With respect to astrology, too, Thagard takes a similarly "thick" approach,
dealing with the specific content and claims of this almost proto-
typical "pseudo-science". Thus, there are no cheap shots here; nor is
there any merely formal analysis, without concrete ballast. What does
show clearly, as well, in the philosophical analysis, is the result of
a half-century of logical reconstruction, in terms of the formal rigor
and logical technique of the analysis and critique. We can no longer
get by with sloppy arguments, or logically loose criticism; and in
these papers, the qualities of logical and analytic rigor, which the
older tradition provides, are in plain evidence, and are deployed in
the treatment of concrete, practical questions.

Finally, what I believe is achieved by the intersection of the
historical and philosophical contributions to this conference is a
recharacterization of science. It is not simply a question of "what

science is", or whether there is such a thing. In a sense, we all know
what science is. It is like the "real world". We all know it's there,
and we know it when we see it, most of the time. We don't make too
many mistakes about that. But there are problematic cases, both historically and at present. And it is these problematic cases, most
sharply brought into question by the normative term "pseudo-science",
that are at issue, and the consideration of which forces us to articulate what we understand by our use of the term science. What we are
trying to do, here, is to construct, reflectively, a critical and useful set of criteria, schemata, case histories, examples, which will
permit us to deal with the extraordinary growth of this particular human activity which we call science, in the last few centuries, and even
more so, in the last few decades. It is no accident that science is so
intriguing. It occupies, and dominates, a good part of our lives and
makes a good deal of our history. Or, one should rather say, *we* make a
good deal of our history by means of our science and its concomitant
technology. What we have come to recognize, in the intersection and
interplay between history and philosophy of science, is how complex and
many-sided, how variable and problematic a human activity science is.
From the historically-informed concrete analysis of the growth of science, and from the philosophically-informed and subtle analysis of the
forms and norms of scientific theory and practice, we arrive at a new
characterization of science: it is neither purely cerebral activity,
nor simply "paper-and-pencil operations", nor a plain reading off of
the "facts of observation" *tout court*, nor the invention of algorithms,
or of algorithmic machines for grinding out results, or manipulating
decision-procedures which we find useful, or interesting. Nor is science just the instrumentality for achieving prescribed and limited results, for we don't know, and can't delimit the results which science
can produce, and so it transcends our particular or local interests in
its instrumental capacities. One of the wonders, and dangers, of science is that when we make scientific progress, we do not only solve a
given scientific problem, but at the same time we create new problems
as well as the potentiality for solving problems which we didn't even
know we had, because we could not have recognized them as problems before. In a sense, science is a problem-generating, as much as a prob-
lem-solving activity. And if this is the case, then it is a much more
complex and many-sided form of human activity than any retrospective
construal of its *achieved* practices can comprehend. It is, in a very

crucial sense, a self-transcending activity. Thus, the epistemological, methodological and ontological presuppositions and approaches
that we take to the study of science are going to have to be different
from the ones that have been taken in the past. And this is not simply because our modes of understanding have changed, but that what it
is we seek to understand, namely science itself, is changing. My
sense of it is that we are at a point of major transition in the understanding of science, as well as of the uses of science. That transition has a great deal to do with the fact that philosophical considerations of a serious sort have been brought to bear upon the historical
reconstruction of science and upon the historiography of science; and
perhaps even more dramatically, an historical awareness has been
brought to bear upon the philosophical analysis or conceptual understanding of science. This conference itself, I believe, is one example
of this important and recent interaction, and its tripartite theme is
an apt expression of the range of considerations which need to be inte-
grated, in this recharacterization of the scientific enterprise.

There is a great deal more I could say, but I think I will save
it for my second life.

FOOTNOTES

1 "The Relation Between Philosophy of Science and History of Science",
   in R. S. Cohen , P. Feyerabend and M. Wartofsky, eds., *Essays in
   Memory of Imre Lakatos*, Dordrecht: D. Reidel, 1976, pp. 718-737.